THE SCIENCE OF MIDDLE-EARTH

HENRY GEE

One of the greatest achievements of Tolkien's *The Lord of the Rings* is its insistent sense of reality; if it is fantasy it is of a very strange sort. Yet equally its genius lies in the wonderful, a parallel world where the marvelous is still possible. Have you ever wondered how Frodo's mithril-coat deflected the deadly blow of the Orc-chieftain in Moria, how it was by naked eye Legolas could without effort count the number of Riders of Rohan across five leagues, or how Galadriel's gift held the light of Eärendil's star? In this engrossing book, Henry Gee ingeniously explores how the marvelous remains marvelous, but not fantastic, and how scientific precepts can reveal some of the wonders of Middle-earth. And in doing so he questions how it is we can grasp the beauty and wonder of Middle-earth, but fail when it comes to conveying the extraordinary nature of science.

—Simon Conway Morris
Cambridge University

How on Middle-earth can the magical world of Frodo and Gandalf teach us anything about the workings of science? What can be 'scientific' about a land of rings of power and silmarils, of balrogs and orcs? Fantasist J.R.R. Tolkien may have had a 'horror of the reductionist program of modern science,' but Henry Gee shows that the author of Middle-earth took an approach to his world that is close to true Science: imaginative, exploratory, creative, and consistent.

Furthermore, Gee uses the events and artifacts of *The Lord of the Rings*, *The Hobbit*, *The Silmarillion*, and Tolkien's other works to examine the latest in scientific research. He shows that the One Ring can teach us about higher spatial dimensions; that the fate of the tree-like ents informs us on population biology and extinction; that Tolkien's own professional field of philology mirrors the reconstruction of the evolutionary history of life. Fans of fantasy and science alike will enjoy such interesting, and unexpected, connections as the optics of the eyes of the elf Legolas and the quantum entanglement of the *palantiri*. *The Science of Middle-earth* may not show you how to forge your own magic sword, but it will help bring the fantastic world of modern science to the readers of Tolkien's classic.

— Thomas R. Holtz, Jr., Vertebrate Paleontologist, Department of Geology and Director, Earth, Life & Time Program, University of Maryland

From Star Trek to Discworld to Harry Potter, scientists have picked up popular series to carry their message of critical enquiry, of education, and of wonder. Henry Gee's one is wonderful, full of wonder, and its last chapter explains helpfully and lucidly what we do it for. This book shows how fantasy enlarges science, and does it beautifully.

— Jack Cohen, co-author, *The Science of Discworld* and *Heaven*

How could phenenomena of the natural world produce effects similar to the enchanting devices of Tolkien's Middle-earth? Henry Gee offers answers in *The Science of Middle-earth*, a tour de force of imagination. What kind of actual materials could compose 'mithril' — no alloy of steel would be strong enough, but perhaps an intermetallic of one of the rare-earth metals? If not an intermetallic, how about silly putty? How could real-life palantirs actually trap light and hold three-dimensional images? In biology, Gee highlights important phenomena rarely addressed by the 'science of' genre, invoking for example underground networks of mycorrhizal fungi to explain the communication system of the ents. At his most outrageous, Gee tackles the thorny question of how did all Tolkien's creatures reproduce themselves? Did orcs make baby orcs, or were they all clones? Or were they actually female parthenotes?

Throughout his book, Gee avoids taking his task too seriously, always bearing in mind that Middle-earth remains fantasy, with tremendous power to stimulate inquiry of all kinds.

—Joan L. Slonczewski, microbiologist and science fiction author

If Charles Darwin were J.R.R. Tolkien, who would be his Thomas Huxley? If *The Science of Middle-earth* is any indication, the answer would be Henry Gee, who in this delightfully written primer explicates the scientific plausibility of Legolas's vision, Ent longevity, mithril flexibility (it's "silly putty"), monstrous wingspan, humanoid evolutionary relationships and a host of other Tolkien techniques too easily dismissible as just magic. The emphasis is quite usefully biology and linguistics. Gee's standards are exacting — he explains that we do not quite yet have the science to account for all the powers of the Ring. But my favorite section was a brief exposition of the word *Tookus*.

—Paul Levinson, author of *The Pixel Eye*

Cold Spring Press

P.O. Box 284, Cold Spring Harbor, NY 11724
E-mail: Jopenroad@aol.com

Library of Congress Control No. 2004109832
ISBN 1-59360-023-2
— All Rights Reserved —
Printed in the United States of America

To my Mother and Father

CONTENTS

FOREWORD
by David Brin

We live in an era of both marvels and marvelous contradictions.

Millions live in poverty and fear. And yet, across the globe, the *percentage* of humans who know peace and comfort is far higher than in any other age. Our Earth strains under the burdens we put on her — but environmental consciousness grows more vigorous and well-informed each passing day. Never before have so many people been so educated, or so daunted by the amount they don't know. Indeed, both the benefits and costs of "progress" seem to pile up at accelerating rates, propelling us toward a horizon of unknown consequences.

This kind of irony extends even to the arts. Under universal education, vast numbers of talented people are given opportunities to create. But this makes it harder for any one creator to stand out... to make a giant imprint on our culture. So where do we look when we feel the yearning for figures who seem larger than life? Is it any wonder that we find many heroes in the past?

This book concerns one of the towering eminences of western literature — J.R.R. Tolkien — whose works had immense impact on several generations, in his homeland and all around the world. In 1997, voters in a BBC poll named *The Lord of the Rings* the greatest book of the 20th century. In 1999, Amazon.com customers chose it as the greatest book of the millennium. Then, of course, there were the recent films, directed by Peter Jackson, together composing one of the most lucrative and successful franchises in the history of cinema.

When a cultural phenomenon burgeons so, it has earned a lot more than just money and attention. It also merits *examination*. Any cultural phenomenon — from rock n' roll to Star Wars — that draws the repeated gaze and adoration of millions, is no longer "just a song" or "just a story." People are moved. Affected. Moral

7

lessons are preached or absorbed. So it was in every prior culture, when tribal members drew inspiration, their codes of honor, good and evil, right and wrong, from stories told around the camp fire, the pulpit, the forum. In order to be so transfixing, a great myth *must* have some kind of deeper content.

Indeed, some are not satisfied just to listen, raptly, or to absorb lessons through the skin. Some of us grow curious *about* the mythic undercurrents — the stuff beneath every superficial stage battle.

Some are spurred to take a closer look.

Henry Gee is one who lived his life taking closer looks. As an editor at the prestigious science journal-of-record, *Nature*, he lives day by day amid the surge of new discoveries that rocket us into the future at unprecedented and accelerating velocity. It takes stamina to drink from that fire hose. Selecting and sampling bits of that torrent to share in *Nature* has trained in Henry Gee a sharp and eclectic eye, not only for detail but also for the Big Picture.

He also happens to be a voracious reader — and sometimes writer — of science fiction and fantasy.

In stodgier times, this statement might have been harmful or even libellous to say about a man in his calling. Science fiction and fantasy long carried a stigma, decreed by those who associated imagination with childishness. But this archaic notion has been changing steadily for generations. Albert Einstein sang the praises of imaginative thinking when he described the importance to science of the *gedankenexperiment* or thought-experiment, in which a researcher first plays out mental scenarios, or fictional stories of "what-if," trying them out for plausibility, picking and choosing just a few that deserve further testing by experimentation.

Henry Gee tested the growing acceptance of *gedankenexperimentation* among scientists in a very bold way, back in the year 2000, when *Nature* celebrated the millennium by weekly presenting a different fictional vignette. These short tales of extrapolation — written by a wide range of authors — posed speculations of wonder and science that stretched minds, entertained, and finally buried the notion of a supposed cultural clash between hard-nosed research and imagination. Nobody's rigor was harmed. Some ideas churned.

We're in the adventure together.

That experiment finished, Gee began mulling another of his loves. Fantasy...especially the worlds created by J.R.R. Tolkien.

Might there be — he wondered — a relationship between Tolkien's marvelously intricate world and that of science? Having read *The Lord Of The Rings* once per year ever since he was ten, Gee already had plenty of notions in that regard.

The result is this book — *The Science of Middle-earth* — one of the most clever and insightful explorations of its kind. Herein you will learn about how Tolkien lived and worked awash in the ferment of ideas, during the first half of the 20th Century, a time when old ways were challenged and the road ahead seemed (in the eyes of many) aimed toward a pit of despair. In the context of such times, and privileged to meet the great minds of his era, Tolkien began spinning legends and tales that led to *The Lord of the Rings* phenomenon.

Could this fantasy have currents beneath those we readily see, on the pages of a simple fantasy tale or on the silver screen? Much has already been made of Tolkien's professional fecundity at building alternate languages for his elves and orcs and other races, all based upon the philology and cultural history of Western Europe, as he knew it.

What Henry Gee boldly aims to achieve is a new level of insight... to explore the *scientific* aspects of Tolkien's vivid world.

I like this book. I would not be writing the foreword otherwise.

This doesn't necessarily mean that I agree with everything contained herein. Henry Gee and I have argued amiably about many aspects, in fact. Interesting aspects, so let me comment.

First, let me avow that I deem Tolkien's trilogy to be one of the finest works of literary universe-building, with a lovingly textured internal consistency that's excelled only by J.R.R.T.'s penchant for crafting 'lost' dialects. LOTR opened the door to a vast popular eruption of heroic fantasy, setting up many others who followed with exacting devotion to his masterful architecture, scrupulously copying the rhythms, ambience and formulas that worked so well.

Indeed, the popularity of this formula is deeply thought-provoking. Millions of people who live in a time of genuine miracles — in which the great-grandchildren of illiterate peasants may routinely fly through the sky, roam the Internet, view far-off worlds and elect their own leaders — slip into delighted wonder at the notion of a wizard hitchhiking a ride from an eagle. Many even find themselves yearning for a society of towering lords and loyal, kowtowing vassals!

In a rather well-known... some might say infamous... essay about Tolkien, I have contended that he was among the most perfect, noble, eloquent and honest promoters of a world-view that I deeply oppose — the *romantic worldview*. Whether or not Henry Gee is right that JRRT shows scientific insight and awareness on the pages of *The Lord of the Rings* (judge for yourself!), it is also worth pondering how two very different worldviews perceive *time* and the notion of *progress*. Since my own passionate interest in these matters contributed to my being asked to write this foreword, let me take just a paragraph or two to explain.

Why do so many films and tales and legends focus on kings and wizards and feudal social orders? I contend it is because we spent many, many more generations under feudal or tribal chiefdoms than the recent sliver of time that democracy held sway. We are — deep down — *used to* the symbols and emotions associated with the old pyramidal hierarchies, in which inherited elites ruled over worshipful peasants - ignorant but loyal - teeming and toiling below. There is something comforting about the image of a Good King, helped by wise, all-knowing wizards or priests.

Anyway, the romantic world view hews to a remarkably consistent set of checklist points. Nostalgia (any golden age is portrayed in the past, never the future). A preference for the pastoral over the urban. For personal loyalties, rather than abstractions (like a Constitution). For a right to rule that is inherited and virtue that is inherent, by right of blood. Evil, too, is inherent to the villain's quality of being. *Crafts* are learned by apprenticeship, as opposed to the more recent notion of *professions* learned at universities. The keepers of knowledge *keep* their knowledge, rather than publishing it openly. Indeed, romanticism celebrates mystery.

There is an opposing world view, one that takes a diametrically opposite position on all counts. Proper fealty is owed to constitutions and laws that were deliberated by elected delegates. Universities and universal education spread knowledge to all who would reach for it. Societies should be "diamond-shaped" rather than pyramidal (the educated and well-off should outnumber the ignorant or poor.) Nobody should get their position by right of birth alone, and claims of inherent righteousness or superiority are suspect. Above all, progress is seen as possible. Not inevitable or uncriticized. But possible. Our children might be better than us. If we are wise.

I have called this the "Enlightenment" view because it grew to full life only in the last couple of centuries, viewing time not as an enemy but as the friend of a humanity that is capable of improvement. Under this view, any "golden age" might be built in the future. Again, if we are wise.

Now clearly J. R. R. Tolkien adhered to one side of this divide, without notable exception. Indeed, he often referred to parts of this division and made no pretensions otherwise. And let me emphasize that *within the context of romanticism*, I have always deemed Tolkien to be a deeply thoughtful, moral, decent and insightful author. Unlike many other romantics, he seemed fully capable of self-appraisal and noticing flaws in the shining armor — for example the deep errors committed historically by his High Elves. Elsewhere I give JRRT much credit for this. I can think of no romantic who I respect more.

Indeed, one of the delights of this book, by Henry Gee, is the chance to see yet another layer in the subtlety of Tolkien's rich mind. For though he was suspicious of progress and the future, he nevertheless clearly reveled in the intellectual excitement that surrounded him. The new discoveries of science were not alien things. With fascination, he seems to have taken a real effort to weave them into *The Lord of the Rings* ... at least so far as they would fit into the grand architecture of elvish lore.

If you are as fascinated as I am by the way ideas churn through the mind of a genius, and how that genius can follow unexpected paths, read on.

David Brin is a scientist, public speaker, and author. A 1998 movie, directed by Kevin Costner, was loosely based on *The Postman*. His fifteen novels, including *New York Times* Bestsellers and winners of the Hugo and Nebula awards, have been translated into more than twenty languages. Brin's non-fiction book — *The Transparent Society: Will Technology Make Us Choose Between Freedom and Privacy?* — won the Freedom of Speech Award of the American Library Association. To learn more about his current work, visit his website at www.davidbrin.com.

AUTHOR'S NOTE

I first came to Middle-earth at around the age of eight, when *The Hobbit* was read to my classmates and me. The reader was almost certainly a lady called Mrs. Elias who made an impression by virtue of her temper and her Porridge Gun, a mythical weapon like a blunderbuss that she would spend whole lessons describing in intricate detail (with diagrams.) My enthusiasm for *The Hobbit* was such that my parents bought me my own copy, from which they were forced to read during my childhood illnesses, engendering in them a loathing for Middle-earth that endures to this day. To their credit, this did not stop them buying me my first copy of *The Lord of the Rings* when I was ten.

I read *The Lord of the Rings* about once a year until I was twenty-five, when I put it aside for a long while — fifteen years, as it turned out. My enthusiasm was rekindled by Peter Jackson's evocative screen adaptation. Picking up the threads where I had left them, I discovered that many things that had gone on in my absence. One of these was the publication of the dozen volumes of *The History of Middle-earth*, edited with extensive notes and commentary by Christopher Tolkien, J.R.R. Tolkien's son and literary executor. This collection of Tolkien's unpublished works, including draft versions of *The Silmarillion* and *The Lord of the Rings*, as well as many works never before seen, provides a revealing insight into Tolkien's own conception of his creation. As I found when preparing this book, *The History of Middle-earth* shows how an increasingly scientific sensibility about such questions as the origin of Orcs had raised doubts in Tolkien's own mind about the consistency of his own legendarium: doubts never made clear until long after his death, but which can now be examined.

Another new phenomenon that had grown up since I last read Tolkien was the World Wide Web. During my reacquaintance with *The Lord of the Rings* I found my way to www.TheOneRing.net (or 'TORn'), the leading Tolkien-related fan site on the web, and it wasn't long before I became actively involved, writing an occasional science column for the site. Some of these columns formed the germs of material in this book. The staff at TORn had produced a book, *The People's Guide to J.R.R. Tolkien*, and kindly introduced me to the publisher, Jonathan Stein, at Cold Spring Press, who added his own comments and suggestions. This book is the result.

This book might seem like another work of Tolkien criticism, and so it is — but one which takes an entirely new and scientific view of Tolkien's work, allowing me to venture, I hope, into hitherto completely unexplored critical territory. After a justification of the application of scientific view to Tolkien, you will see a selection of loosely connected essays on various aspects of Tolkien's legendarium, showing how a modest amount of scientific understanding can enrich the appreciation of Tolkien's works as literature. I close with an essay entitled 'Science and Fantasy' in which I demonstrate that Tolkien's own world view was closer to the true spirit of science than that held by many who propose to promote the public understanding of science.

Some notes about notes. In what follows, I shall assume that the reader is familiar with *The Hobbit*, *The Lord of the Rings*, the published version of *The Silmarillion* and to a lesser degree *Unfinished Tales*, but has a more casual acquaintance with *The History of Middle-earth* and with science. Mention of particular episodes in *The Hobbit* or *The Lord of the Rings* is generally made without any formal reference. Where I have felt a citation to be necessary, I refer to *The Lord of the Rings* by the abbreviation '*Rings*,' with an indication either of the book (not volume) and chapter wherein the relevant passage may be found. Thus '*Rings* III, 2' refers to the chapter 'The Riders of Rohan' in *The Two Towers*; or of the relevant Appendix, so '*Rings* F' refers to *The Lord of the Rings*, Appendix F.

I refer to material in *The History of Middle-earth* by the abbreviation '*HOME*' with the volume number. For example, '*HOME* XII' refers to *The History of Middle-earth Volume XII: The Peoples of Middle-earth*. Where appropriate I refer to particular identifiable sections of these volumes, for example '*The Notion Club Papers*, published in *HOME* IX.'

I refer to *Unfinished Tales* and the published version of *The Silmarillion* by section (for example *Ainulindalë, Valaquenta*) or by chapter (if in the *Quenta Silmarillion*.)

AUTHOR'S NOTE

References to material in *The Letters of J.R.R. Tolkien*, edited by Christopher Tolkien with the assistance of Humphrey Carpenter (henceforth '*Letters*') are cited by the numbered item of correspondence in which the material appears, rather than the page (for example, *Letters* 266 refers to the letter numbered 266, not to page 266).

Tolkien's essays *The Monsters and the Critics* and *On Fairy Stories* can now be found in *The Monsters and the Critics and Other Essays*, edited by Christopher Tolkien (1983). His scholarly work *Finn and Hengest: The Fragment and the Episode*, edited by Alan Bliss (1982), is also available in paperback.

Given that many editions of Tolkien's work exist, and that readers will almost certainly have different editions from my own, I felt that giving citations to page numbers in most of these works would be confusing. In any case, the comprehensive indexes for all these books should allow the reader to pinpoint particular references with ease.

Turning to Tolkien-related work by others, I refer to Humphrey Carpenter's *J.R.R. Tolkien: A Biography* (first published by George Allen and Unwin in 1977) simply as '*Biography;*' Tom Shippey's books *The Road to Middle-earth* (HarperCollins, 1982, revised 2003) and *J.R.R. Tolkien: Author of the Century* (HarperCollins, 2000); Verlyn Flieger's *Splintered Light: Logos and Language in Tolkien's World* (Kent, Ohio: Kent State University Press, revised edition, 2002) and *The People's Guide to J.R.R. Tolkien: Essays & Reflections on Middle-earth from TheOneRing.net* (Cold Spring Press, 2003) either informally and as a whole, or by chapter where I consider a more specific citation necessary.

References to all other works are given as fully as seems appropriate, given that publication details vary from place to place, and that this book is meant as an entertainment more than a scholarly tome. To avoid continual page-turning I have kept extraneous notes brief, although this has not always proved possible or desirable, and additional material of interest — including most citations to scientific sources — has been placed in a section of notes at the end of the book. These are indicated in the text by numerical superscripts.

I have not tried to be comprehensive either in the material I have discussed here, or with the number of examples from the thousands of pages of Tolkien's works I could have chosen. I have selected only those examples I needed to make a point: to have gone any further would have produced a far weightier tome than this. Besides, it would have left me with less scope, perhaps, to think of more things to write about in the future. And if you, the reader, can think of ways I could have addressed the issues discussed in these pages, identify mistakes, or think of things I haven't covered, I should welcome your views, to which full credit will be given, should the opportunity arise.

I should add that this book is completely unofficial. Although I am an occasional guest columnist for TheOneRing.net, I have no connection with the Tolkien Estate, Tolkien Enterprises, New Line Cinema or any other organization that has a claim on the works of J.R.R. Tolkien, Christopher Tolkien, or Humphrey Carpenter. I have not attempted to contact any of these bodies for their approval or endorsement, so you should take what follows as entirely unauthorized.

Writing this book would have been a pleasure even had I written it in a state of complete isolation, but the course has been enriched by interactions with many others. I thank my agent Jill Grinberg and Jon Stein and his colleagues at Cold Spring Press for making it a reality. I thank Luis V. Rey for painting the wonderful picture on the cover. I thank the staff of TheOneRing.net, especially Quickbeam and Turgon, who graciously allowed me to write for the site, encouraged me to use my articles there as testing grounds for material in this book, and who made many interesting comments and suggestions. 'O for the Wings of a Balrog' and 'Holes in the Ground' grew out of suggestions causally tossed aside by Turgon like so many pearls on the strand of Eldamar.

I owe a special debt to Edmund Gerstner, Christopher Surridge and Karl Ziemelis, collectively TolTec – the informal Tolkien Technology Research Group, made up of colleagues with whom I discussed many of the scientific issues raised by Tolkien's fiction. 'Inventing the Orcs', 'Armies of Darkness', 'Of Mithril', 'The Laboratory of Fëanor' and 'The One Ring' owe much to this collective effort. Simon Conway Morris, Eichling and Kristine M. Larsen provided many detailed and thought-provoking comments on the text: I am particularly grateful to Eichling for sharing her first-hand knowledge of the mechanics of mail- and plate

AUTHOR'S NOTE

armor, which was especially helpful in 'Of *Mithril*.' Catherine Cassidy, Craig Davidson, Debbie Hazell, Clare Ireland, Mark Isaak, Tony Kerstein, Sandy Knapp, Sandra Last, Brian Mason, Kathryn Phillips, David G. Roberts, Adam Rutherford, Joshua Siary, Yarrow Wood and Emre Yigit provided helpful comments, insights and encouragement.

I thank my wife Penny for her patience (she says I spend hours in the study just tolkien to myself), and my children Phoebe and Rachel for being in the vanguard of the next generation of Tolkien fandom. I have had the pleasure of reading *The Hobbit* to them from the same copy that enriched my own childhood. I dedicate this book to my parents, Rita and Tony Gee, for their forbearance in indulging an early passion that they could not share.

And I thank Mrs. Elias for starting me off. If you're out there, Mrs. Elias, do let me know. I'd like to know if you've perfected the Porridge Gun.

ABOUT J.R.R. TOLKIEN

John Ronald Reuel Tolkien was born on January 3, 1892, in Bloemfontein, South Africa, where his father was a bank manager. In 1895 his mother took him and his younger brother on home leave to England. After Tolkien's father died the following year, the family remained in England and Tolkien grew up among his mother's kin, in the West Midlands, now the suburbs of Birmingham. In 1900, Tolkien's mother was received into the Roman Catholic Church. She died four years later from complications arising from diabetes, aged just 34. Her religious views resulted in estrangement from her Protestant family: this, and her early death, was something Tolkien never forgot or wholly forgave, and deepened his own Catholic faith.

Now orphaned, the Tolkien brothers lived in a succession of lodgings while being educated at King Edward's School in Birmingham. In 1910 Tolkien went up to Oxford. After graduation in 1915, Tolkien joined the army, married his fiancée Edith Bratt, and in 1916 served on the Somme, returning to England that same year with trench fever. During this period he began work on *The Book of Lost Tales*, the germ of *The Silmarillion*.

After the Armistice in 1918 the Tolkien family returned to Oxford where Tolkien worked on the New English Dictionary. In 1920 he moved to the University of Leeds as Reader in English Language, becoming Professor in 1924. The next year he returned to Oxford once more, this time as Rawlinson and Bosworth Professor of Anglo-Saxon, and soon befriended C. S. Lewis. Their friendship formed the nucleus of an informal but long-lasting literary association known as 'The Inklings.'

In public, Tolkien lectured on *Beowulf*; in private he was working on *The Silmarillion* and, as an amusement for his children, *The Hobbit*, which was published in 1937. On the suggestion of his publisher, Tolkien immediately started work on a sequel. This took twelve years and another war to complete: *The Lord of the Rings* was finally finished in 1949, and not fully published until 1955, when Tolkien was sixty-three. Tolkien died in Bournemouth on September 2, 1973, aged eighty-one. A full account of Tolkien's life can be found in Humphrey Carpenter's *Biography*.

INTRODUCTION

The increasing technological sophistication of our society is matched by a decline in the number of people willing to learn about science and engineering. Everyone wants to use the latest gadgets — but there are many more people who care about the social and political implications of technology than willing to understand how it works. Such ignorance is a function of the degree to which technology has become commonplace.

Not so long ago, the products of high technology were unusual, cumbersome, expensive, difficult to use and frequently went wrong. Therefore it was useful to have a working knowledge of the internal workings of a device, if only so that you could give a creditable account of its faults to a repairman. Nowadays, we expect technology to be completely reliable, as easy to use as breathing, and so cheap that in the unlikely event that it *does* go wrong, replacing an outmoded gadget with a newer model is an affordable, perhaps even expected option. The remarkable thing is not that the technology is amazing, but that it usually meets our expectations.

As a result of this convenience, products of technology are as ubiquitous as hamburgers and shoes, and we treat them the same way, as adornments to our lifestyle. In other words, we take the modern magic of technology for granted, and can see no reason why we should learn the technical niceties of, say, wireless internet technology, any more than we should be compelled, as if for our own betterment, to visit shoe factories. Because the details of technology are in themselves uninteresting or even baffling to most people, we are more concerned with issues raised by the *use* of technology, such the problems of recycling large piles of the discarded products of technology that cheapness and fashion have made obsolete, or its effects on health (can mobile phones cause brain tumors? How often do we need to take screen breaks at work?) In short, there is no more compelling need for most people to understand the inner workings of technology than to indulge in any other pastime we might choose.

Scientists and politicians, however, see a problem: that without trained scientists and engineers to invent things and create wealth, our industry will soon cease to be competitive, and we shall all lose out to the industrial might of countries who might not have our interests at heart. This anxiety has led to strenuous and increasingly desperate efforts to educate a seemingly uncaring public about science, as if time spent doing other things was time wasted, and more profitably employed in learning about evolution or gravity. This aggressive promotion of science is flawed, not least because many of its provisions are, strange as it seems, antithetical to the pursuit of science.

I discuss this particular issue at the end of the book. For the moment I'd like to advance the idea that in a post-industrial society, knowledge of science should not be seen as a prerequisite to the creation of wealth so much as a part of our culture in general, like the appreciation of music, literature or good wine, and not something that stands in utilitarian opposition to what is generally regarded as 'culture.' Now that the industrial revolution has been and gone, an appreciation of science in the modern world should, instead, resemble what it might have been in the Age of Enlightenment of the eighteenth century, when knowledge of the natural world stood alongside the classics and scripture in the intellectual furniture of any well-educated person. To pursue science as a vehicle for wealth creation alone is as unappealing as any other chore, which explains why young people would rather learn about drama than physics.

I propose instead that our understanding of science should rest on the same sensibility of leisure with which we appreciate art, and, through this, seek to learn more: after all, pleasure — rather than profit — is the motivation that drives young people even outside school hours to know everything there is to know about dinosaurs, or the Solar System. Innate curiosity, not some prescription of knowledge drawn up by others who presume to know what is best for us, is what drives small children to the extent that they can pronounce words such as 'triceratops' while still in diapers.

To pursue this idea further, it should be evident that our understanding of different branches of inquiry becomes more enjoyable if it is interactive, that is, if we can build bridges — make connections — between separate strands of thought. Our appreciation of any piece of music is heightened if we know who composed it, something about their life, and why they wrote the piece as they did. For example, the innovation of Beethoven's final string quartets is put into perspective once we know that the composer was deaf when he wrote them, and having come to terms with that fact, became free to create new and challenging soundscapes. Our knowledge of Beethoven's life enriches the music we hear.

Science writers have been generally slow to pick up on this trend towards synthesis, that is, the portrayal of science as one cultural mode among many, and not a thing apart. The signs, however, have been clear to see in the bestseller lists. *A Brief History of Time*, Stephen Hawking's uncompromising introduction to cosmology, may have sold on the circumstances of the author's life as much as on the subject matter. Dava Sobel's *Longitude*, about one man's quest to make a reliable chronometer, succeeds because it is a good story about an interesting period of history, and its contribution to the public understanding of cartography or clock-making is largely incidental. The lesson is clear. Writing about science succeeds if it is married to a mode more accessible to the culture in general. Science writing for its own sake is no more than a genre, with its own formula and its own fans, but less likely to penetrate the general market on its own merits.

Genre writing of all kinds — westerns, romance, science fiction, fantasy, horror and, to a lesser extent, crime stories — are conventionally sidestepped by critics of mainstream literature, and the same is true for science writing unless there is some explicitly non-scientific angle with which it might be promoted. Because of this, cross-fertilization between genre writing and branches of so-called 'high' culture has been hard to achieve. Fans and serious critics of genre fiction often profess puzzlement at its exclusion from the cultural high table. In *The Road to Middle-earth*,

Tom Shippey poses the question of why academics concentrate on a relatively small part of literary endeavor and ignore the vast canon of literature that millions of ordinary people read purely for enjoyment. The reason is not hard to see — because genre fiction is seen as popular and demotic, it is regarded as insufficiently refined to be worthy of study. By concentrating on 'serious' literature that relatively few people actually read, critics are affirming their own exclusivity as self-appointed secular priests whose judgment of what is 'good' carries weight only because we (who wish to join this exclusive club and feel ourselves to be intellectually superior) listen to their pronouncements.

Although scientific themes have made some inroads into serious literature, instances are notable for their rarity, and when literary authors stray into genre territory, they are at pains to deny it. When literary authors write science fiction, they tend to call it something else, such as 'magical realism.'

Science makes more explicit appearances when associated with popular culture, as manifested by a recent rash of books on the scientific aspects of successful fantasy and science-fiction books and TV series. To cite just some examples — there's *The Real Science Behind the X-Files* by Anne Simon; *The Physics of Star Trek* by Lawrence M. Krauss; *The Science of Harry Potter* by Roger Highfield; *The Science of Philip Pullman's 'His Dark Materials'* by John Gribbin and Mary Gribbin, and two volumes of *The Science of Discworld* by Terry Pratchett, Ian Stewart and Jack Cohen – with a third promised. In this context, the hit TV series *Walking With Dinosaurs* (with accompanying books, DVDs, videos, and merchandise) might be resold as *The Science of Jurassic Park*.

Of all the works of genre fiction, none is more enduringly popular than *The Lord of the Rings*. However, as far as I am aware, nobody has written a *Science of Middle-earth*. At first sight, there appears to be a good reason why: that is, Tolkien's well-advertised loathing of science and technology. Embraced by environmentalists, lionized by tree-huggers, almost a patron saint of the technophobic, Tolkien is perhaps the very last person whose work you'd expect to be the subject of scientific treatment. Tolkien's work is everywhere littered with paeans to rustic holism and a horror of the reductionist program of modern science, in which the way to understand how a thing works is, first, to break it into pieces — a scheme that is explicitly damned as foolish by Gandalf in his criticism of his rival Saruman (*Rings* II,2).

In this book I shall show that Tolkien's stance on science is not as simple as it first appears. Examination of Tolkien's works show that he prized science as the acquisition of knowledge. What he deplored was its use as an instrument of domination. Tolkien's legendarium is based on the revolt and subsequent long repentance of the Noldor, the most 'scientific' of all the kindreds of the Elves, devoted to the exploration and understanding of the material world. Their chief was Fëanor, the greatest of all the Elves. It was Fëanor who created the Silmarils over which the first wars were fought. It was the descendants of Fëanor who forged the Rings of Power and who fought Sauron in the Second Age. The last of the Noldor is Galadriel, the beautiful, near-omniscient and rather frightening Lady of Lothlórien. To be sure, Tolkien was fond of the rustic life personified by the typically unadventurous Hobbits, but Middle-earth would have been a dull place indeed were it not for the Noldor and the disasters that followed their relentless quest for knowledge. Without the Noldor there would have been no *Silmarillion*, no *Lord of the Rings* – in short, no story.

Tolkien's point is not that knowledge is a bad thing of itself, but that it is perilous to acquire knowledge without either seeking to understand the context in which that knowledge must be placed or, recklessly, without counting the cost, and in these respects many scientists would agree. The interplay between discovery and responsibility — between creation and subcreation; between blue-skies research for the fun of finding out, and research directed towards more mercenary ends — is a major theme in Tolkien's fiction, as it is a preoccupation of scientists involved in such controversial areas as the biology of infertility or the genetic modification of living organisms. For this reason alone it is worth approaching *The Lord of the Rings* from a scientific viewpoint.

But it is not the only reason. An examination of Tolkien's letters and unpublished works such as *The Notion Club Papers* (published in *HOME* IX) shows that Tolkien was acutely sensitive to, and conversant with, various themes in the literature of science fiction, a genre that was assuming its modern form during the 1930s and early 1940s — Tolkien's most productive period: he was not quite the Luddite he claimed to be. Perhaps more deeply,

Tolkien's professional work as a philologist was scientific, by his own admission (*Letters* 163), and yet nobody, as far as I am aware, has remarked on just how scientific this discipline really is — mainly because, since Tolkien's day, philology has moved away from departments of English Literature, and therefore from mainstream criticism. In the meantime, comparative linguistics has become associated with science, in particular genetics and evolutionary biology, with which it has many parallels. Because of this academic migration, modern research in the kind of philology in which Tolkien might have been interested has largely disappeared below the radar of literary critics, who, after all, remain largely unschooled in science.

This book is not a how-it-works guide to the Ring or Elvish 'magic.' To be sure, you'll find things of that sort in what follows, but to concentrate solely on the purely mechanical would be a disservice to a work that transcends genre fiction. In contrast, my reason for writing this book is to show how the application of scientific knowledge can add new and exciting perspectives to our understanding of Tolkien's work as *literature*, deepening our understanding of the complexities of the plot and the motivations of the characters. A little knowledge of science should be an essential part of the training of any literary critic, just as a little literature is vital for the training of any scientist or, indeed, any human being.

1. SPACE, TIME, AND TOLKIEN

Fate, destiny, Elves, Dwarves, the extinction of species — all are familiar themes from fairy-tales, but they find contemporary articulation in science fiction and fantasy. Few knew more about the mechanism of myth than Tolkien, and he was keenly aware of the role of science in the more contemporary forms of literature in which he and his peers, particularly C.S. Lewis, were interested.

Lewis and Tolkien were as aware as any modern science-fiction critic that 'hard' science fiction (that is, science fiction in which science and technology are depicted with realism) differs from fantasy only in their props. Gandalf's staff is functionally equivalent to Obi-Wan Kenobi's light saber, just as Vulcans and Klingons are Elves and Orcs recast; the *palantír* fulfills the same function as the 'ansible' as found in Orson Scott Card's *Ender's Game*, or indeed that of any other 'deep space' communicator. By this reading, *The Lord of the Rings* belongs firmly in the canon of science fiction alongside Lewis's *Out of the Silent Planet* and every space-opera ever written.

This may explain Naomi Mitchison's jacket blurb on the first edition of *The Lord of the Rings* in which she described it as 'super science fiction.' This categorization seems jarring at first, until you realize that Tolkien himself identified very strongly with science fiction. He admired the work of Isaac Asimov and particularly in later years read much science fiction and fantasy (*Letters* 26, 294 and 297). Just how deeply Tolkien was immersed in science fiction during its 'Golden Age' of pulp publication in the 1930s and 1940s is made clear in the characteristically unfinished story *The Notion Club Papers* (now published in *HOME* IX).

Tolkien started *The Notion Club Papers* as a kind of breather in the mid-1940s, between finishing *The Two Towers* and starting in earnest on what became *The Return of the King*. It is conceived as the fragmentary record, found in mysterious circumstances, of the meetings of an assortment of varyingly eccentric Oxford academics. Their gatherings are purportedly held in the 1980s and 1990s, their records discovered only in the early 21st century: but the whole exercise is then suspected to be a scam, a futuristic fiction written much earlier, in the 1940s. I confess to being rather fond of this kind of cod-literary bluff and double-bluff, which is more typical of authors such as Jorge Luis Borges than Tolkien.

Indeed, *The Notion Club Papers* has a lot in common with Borges' essay *Tlön, Uqbar, Orbis Tertius*, in which a secret academic society seeks to invent a world in such detail that people actually start to believe it exists, and so it does. Indeed, Tolkien later raised the idea of the formation of an invented world as a kind of game, perhaps contrived by a committee of experts (*Letters* 154). This could almost be a comment on the process of subcreation in which inventors of fantasy landscapes such as Middle-earth seek to make them so real that people can, in Tolkien's terms 'suspend disbelief' while they 'visit' them.

This idea was taken to a hilarious extreme in *Madame Bovary, C'Est Moi*, a vignette by science-fiction author Dan Simmons[1] in which people can 'teleport' to universes created by authors of fiction, provided that the universes are sufficiently believable and self-consistent that they generate an 'entangled-pair consciousness wavefront' in the structure of space time in which human consciousness can become enmeshed. Once this is discovered, people teleport to the Universes of *King Lear*, *Hamlet* or a host of others. The inventor of the teleportation technique lives as a character in the world of Flaubert's *Madame Bovary*. This piece seems, at first, to be no more than an enjoyable romp on quantum teleportation, one of the latest and most intriguing scientific results.[2] However, below the surface it is evident that Simmons asks serious questions about the nature of our own reality — owing much to Borgesian stories such as *The Circular Ruins*.

Beneath its initial Borgesian conceit, *The Notion Club Papers* is a contrived lampoon of the Inklings, the informal literary drinking club of which Tolkien and C.S. Lewis were the best-known members. At least, that's the way it starts. It ends as a reworking of Tolkien's own version of the Atlantis legend, a persistent fixation that came to him in recurrent nightmares in which landscapes were drowned beneath a towering wave (*Letters* 163).[3] Many years before *The Notion Club Papers*, Lewis and Tolkien had set each other a challenge in which Lewis would write a story on space travel, whereas Tolkien would do something on time travel. Lewis, with characteristic energy, produced *Out of the Silent Planet* (1938), which led to two published sequels — *Perelandra* (1943) and *That Hideous Strength* (1945). Lewis's science fiction trilogy has become a classic of the genre — Tolkien's effort, however, foundered after a few pages and never saw the light of day. It was called *The Lost Road* (now published in *HOME* V).

Like much else in Tolkien, the story of *The Lost Road* is based on a philological trick, in which pairs of fathers and sons through time have names that can be translated, respectively, as 'Bliss Friend' and 'Elf Friend.' Because of this, a modern 'Edwin' and 'Alwin' find themselves in a kind of rapport across time with a Saxon Eadwine and Aelfwine (the latter a character recycled from the even earlier *Book of Lost Tales*), a pair of Lombardic chieftains whose names had similar derivations, and so back to the original casting of Amandil and Elendil, the 'faithful' who escaped the wrack of Numenor. Other fragments of northern legend are brought in along the way, most notably Old English poems such as *The Seafarer* and the persistent myth of King Sheave (which turns up in *Beowulf*), in which a boat comes from the West bearing a beautiful and royal child who grows up to bring prosperity to the hither lands (see Christopher Tolkien's own notes on the King Sheave myth and much else in *HOME* V.)

In Tolkien's mind, this story would be all that would have remained in primitive English traditions of the coming of the ships of Numenor to Middle-earth. After a few chapters, *The Lost Road* became enmeshed in philological distractions, and Tolkien abandoned it. Years later he professed himself ashamed of it (*Letters* 105) even though from its admittedly scruffy soil grew the later unpublished work on *The Fall of Numenor* (to which *The Notion Club*

Papers is closely related); the mighty *Akallabêth* (published in *The Silmarillion*); and indeed the entire Second and Third Ages on which *The Lord of the Rings* is based.

Like *The Lost Road*, *The Notion Club Papers* starts with good intentions as a story, but ends up in precisely the same quagmire in which Numenor, Elvish, Old English, King Sheave, Elendil, Sauron, Aelfwine, the then-emerging language of Adunaic and even the mythical Voyage of St. Brendan find themselves entwined in a thoroughly overcooked stew. And so the story peters out in muscle-bound exhaustion: Tolkien abandoned the *The Notion Club Papers* and took up *The Return of the King* where he had left it.

The difference between *The Lost Road* and *The Notion Club Papers* concerns the setting in which all these mythopoeic and linguistic threads are woven. The setting of *The Notion Club Papers*, as a lively ensemble piece of variously eccentric academics, makes a refreshing and surprisingly successful contrast with the stilted, unrealistic father-and-son two-piece of *The Lost Road*. As a kind of self-referential (and perhaps slightly envious) post-mortem on the experiment that led to *The Lost Road* and Lewis' justifiably more successful novels, the early pages of *Notion Club* focus on donnish discussions of Lewis's space fiction – which led to Tolkien's working title of *Out of the Talkative Planet* for his new work. But apart from this fun if rather self-indulgent chit-chat, the central image of *The Notion Club Papers* is just the same as in *The Lost Road* and all the subsequent work on Numenor – that is, the destruction of an Atlantean civilization and the 'bending' of the Earth, so that the 'straight road' to the blessed realm in the West is vouchsafed only to a few, and then only haltingly, in dreams or visions.

The idea of a 'lost road' to the ancient world of the Elves that can be discovered only by a few, blessed people goes back to Tolkien's earliest writings. In *The Book of Lost Tales* (*HOME* I) Tolkien discusses it in terms of a secret path to a place that can be visited only by children in their dreams. The idea of a secret dreamworld accessible only to children or the adept is a staple of children's fantasy literature, from *Alice's Adventures in Wonderland* to Baum's *The Wizard of Oz*. It is characteristic of Tolkien, however, to suggest that what we now discuss only as nursery lore – along with the Elves themselves – once had a more serious and material existence, and the idea of the 'lost' or 'straight' road to the 'true west' soon outgrew its folksy beginnings.

As in *The Lost Road*, the action in *The Notion Club Papers* involves just two of the company, who find themselves sensitive to the ethereal vibrations of history, and through a kind of trance state recover fragments of ancient languages (Quenya, Sindarin, Old English and Tolkien's new invention, Adunaic.) These gifted members of the Notion Club cause consternation when they vanish for weeks at a time to collect old sea-lore from western parts of Britain, only to turn up at meetings of the club where they interrupt proceedings by staring significantly at the sunset and shrieking lines such as 'Behold! the Eagles of the Lords of the West!' in one of a choice of real and invented languages, before falling into a melodramatic funk. At this point Tolkien starts writing what looks like a (surely unintentional) parody of the overwrought gothic horror of Howard Phillips Lovecraft (1890-1937).

However, there are a few pages in *The Notion Club Papers*, between the Borgesian mind-games at the start and the Lovecraftian morass at the end, in which Tolkien betrays more than passing acquaintance with one of the most important themes in science fiction, that is, the technology of space travel. Given that Tolkien is usually seen as the *doyen* of medievalist, pre-technological fantasy, it may come as a surprise that he knew about what was then known of space and space technology, not to mention concepts such as the theory of relativity, and how these things were depicted in fiction.

In these early passages, the Notion Clubbers talk about how Lewis's characters traveled between planets, comparing Lewis's ethereal means of transport in, say, *Out of the Silent Planet* with H. G. Wells' more concrete devices in stories such as *The Time Machine* (1898) and *First Men In The Moon* (1901).

In the latter, Wells has his spacecraft powered by 'cavorite,' a mineral discovered by his hero, Cavor. The function of cavorite is to act as a source of anti-gravity: the Notion Clubbers, discussing Wells explicitly, argue that for writers to substitute technological-sounding mumbo-jumbo in place of real knowledge is dishonest, and not only that, an affront to the universe as we know it.

In the words of one of the party, Nicholas Guildford, gravity is a fundamental statement of where you are in the universe, and cannot be tricked by inventing a bogus if scientific-sounding name such as 'cavorite.' This objection shows an acute sensitivity to the theory of relativity (according to which, gravity has *precisely* the function of spatial location that Tolkien reads into it). It also displays the same disdain that Einstein famously showed for quantum mechanics: if God does not play dice with the universe, says Einstein, then — continues Tolkien — neither should we.

In general, Tolkien's character Guildford expresses distaste for the abracadabra that science fiction heroes use to get from A to B, noting that in such stories, accounts of the technicalities of journeys in space- and time-travel tales always seem the weakest, acting only to disrupt the momentum of the narrative. Guildford's point is that fictional machinery, if not honestly conceived, is a disservice to the tale as a work of literature, not to mention a distraction from the story. If you cannot give a convincing explanation of how your character gets to his destination, consonant and consistent with what we really know about the behavior of the Universe as we know it, then there is no point dwelling on the technicalities, for such things will undermine the flow of the story. All in all, says Guildford, it would be best to have a wizard wave his wand and have done with it. It may sound arbitrary, but it is no more or less so than scientific-sounding props such as cavorite, and it is at any rate more honest. A magic wand is a more straightforward statement of one's own ignorance, for one is not obliged to wonder how it works, whereas cavorite is a more obvious expression of science, which, as such, might beg further explanation.

The more fundamental point, about the conjuration of technology in defiance of the Universe, finds contemporary echoes in science fiction. In a novel called *The Songs of Distant Earth*, Arthur C. Clarke explores the human colonization of the galaxy, aiming to be as realistic as possible – like Tolkien in *The Notion Club Papers*, to fit the universe as it is, not as we should like it to be. Although Clarke describes some fantastic technologies in his novel, such as the idea that we might harness the stupendous energies of the 'quantum foam' that some physicists think underlies reality, these are all based on real science, or at least informed scientific speculation. For this reason, Clarke pointedly eschews *Star Trek*-style technological faster-than-light travel as very much the kind of device so abhorred by Tolkien, describing it as the kind of mumbo-jumbo that conveniently allows the Great Producer in the Sky to get from one location to another in time for Next Week's Exciting Episode. The drama and poignancy of *The Songs of Distant Earth* result from the use of ever more sophisticated versions of subluminal travel (i.e., travel slower than the speed of light): if you are going to use some unimaginable technology to travel faster than light, or back in time, or between planets, just wave a wand — or dream yourself into it.

During this discussion, Tolkien uses an intriguing word that reveals a great deal about his knowledge of science and science fiction, which turns out to have been far more extensive than headline exposure to H. G. Wells might suggest. That word is 'scientifiction,' and Tolkien uses it three times in *The Notion Club Papers*. Given Tolkien's generally extremely careful choice of words, the use of this word can only be seen as significant.

What we now call 'science fiction' started with the vision of one Hugo Gernsback (1884-1967), who came to the United States from Luxembourg in 1904 and made his way as a publisher of cheap magazines. His title *Amazing Stories*, launched in 1926, was the first English-language magazine devoted to what Gernsback himself called 'scientifiction,' a word he himself coined as the name for the genre. It is an amusing thought that the same pulps that inspired youthful writers such as Isaac Asimov, Frederik Pohl and Arthur C. Clarke might have found their way into the study of an Oxford English professor.

After dismissing *faux*-scientific contraptions such as cavorite, the Notion Clubbers start considering more philosophical methods of travel in time and space, which two of the company eventually use to assume the characters we now know as Amandil and Elendil, and are able witness the downfall of Numenor. Here Tolkien once again demonstrates the extent of his reading, when one of the characters, Ramer, remarks to another, Wilfrid Jeremy — an authority on speculative fiction — that he had become interested in telepathy as a literary device for viewing events remote in time, citing as a source a novel called *Last Men in London* that Jeremy had lent him. *Last Men In London* tells how the final species of Man — living on Neptune two billion years in the future

— communicates telepathically with members of our species, living in 1930s London.

This was a real novel, published in 1932, of which Tolkien must have been aware, and one of the small but remarkable output of William Olaf Stapledon (1886-1950), one of the neglected geniuses of modern literature that mainstream literary criticism unaccountably seems to ignore.

Stapledon had many things in common with Tolkien, and it would be interesting to know whether they ever met or corresponded. Just six years older than Tolkien, Stapledon, like Tolkien, served in World War I (in the Friends' Ambulance Unit, rather than as a soldier); also became a professional academic; and also did most of his best writing as the world slowly slid into World War II.

Last and First Men (1930) — to which *Last Men in London* is a kind of appendix — tells the entire billions-of-years saga of humanity, a tale conceived on the kind of audacious scale that few writers have dared attempt, before or since. If that were not enough, the entire action of *Last and First Men* is told in barely two paragraphs of *Star Maker*, published (like *The Hobbit*) in 1937, on the very eve of war.

Star Maker is less a story than an almost painfully intense vision in which a modern man, tortured by the circumstances of his day, leaves his suburban home one evening to sit on a nearby hillside and look at the stars. Almost without his own volition, he is caught up in what he calls a 'hawk flight of imagination' — just the kind of honest means of space travel that commended itself to the Notion Club — that carries him to the ends of the cosmos, where he finds that the very stars themselves are sentient and, as part of a community of disembodied spirits, he gets to meet the Creator. And even as the Creator makes each universe more perfect than the last, there is still darkness and suffering beyond imagining or comprehension, which we must do our best to combat before night falls. Stapledon, like Galadriel, is concerned with fighting the long defeat (*Rings* II,7), albeit on a stupefyingly grand scale.

From all this it is clear that Tolkien, far from shunning the concepts of science, had more than a passing knowledge of the scientific concepts of the day, and understood the degree to which these concepts had become established in reality such that we could legitimately use them in stories. In this sensitivity, Tolkien was far from being a wide-eyed bucolic fantasist, and was, in this respect at least, as hard-nosed a writer of so-called 'hard' science fiction as Asimov or Clarke. Tolkien realized that if you are going to mount a cogent critique of the uses and abuses of science, you have a duty to know all about it — otherwise, you would be doing nothing more effective than throwing rocks from the sidelines.

Tolkien's knowledge of science makes the moral pitfalls facing his 'scientists' — the jealous Thorin and his Arkenstone in *The Hobbit*; the ring-making Elves of Hollin; Fëanor and his Silmarils; Saruman and his industrial urges; Sauron and his thirst for absolute power and control; Aulë and his creation of the Dwarves in the *Silmarillion*, even Morgoth — all the more interesting, and their choices more complex. For none of Tolkien's main characters is damned simply by his association with science, only by the choices each one makes about how the knowledge gained by their scientific researches is to be appropriated or shared.

It is this point — about science, more than mere tree-huggery — that gives Tolkien's work such a contemporary air, for it resonates with the subtleties of current arguments about technology. The argument about genetically modified crops is not that they are intrinsically right or wrong, but about the degree to which the technology is controlled by particular interests; the argument about drugs that fight HIV-1, the virus that causes AIDS, is not about whether HIV-1 really is the cause of AIDS, but about how these drugs can be distributed in a way that is affordable both to the people who need them and the companies that develop them; the argument about *in-vitro* fertilization and similar methods of assisted conception is not just about morality, but on how and whether such technology can and should be made available to all who feel they need them at the expense of public resources. Contrary to popular belief, Tolkien did not damn science unreservedly. He had taken time to understand its workings, and thought very carefully about how scientific concepts might best be presented in fiction.

2. INSIDE LANGUAGE

That Tolkien had a great deal of respect for science should come as no surprise, in that he considered himself to be a scientist (*Letters* 163, 257) — a scientist of language and words. This was not an idle or fanciful label. Philology, Tolkien's field of study as a professional academic, and on which all his fiction is based, is much more a science than many other activities that bear the label more prominently.

The Oxford English Dictionary defines 'philology' first as the love of literature and learning in its widest sense, but notes that the use of the word to mean the study of the structure and development of language — in short, the science of language — is the more common. The OED also notes that this second, perhaps more familiar usage is really a narrower subset of the first, as if a field of enquiry that once roamed the widest plains of knowledge had become restricted to a dry and straitened creek.

Changing fashions in the study of language have mirrored the change in the usage of the word itself. As Tom Shippey recalls in *The Road to Middle-earth*, philology was in retreat as a part of university English curricula even when Tolkien began to teach it, in the 1920s.

At the time, English syllabuses were divided into two distinct halves. 'English Literature,' or 'Lit.' concentrated on criticism of literature from the time of Chaucer onwards. 'English Language' or 'Lang.,' the other branch of the English syllabus, looked at the evolution of the English language from its ancient origins. English Language, including philology, was objective, comparative, and rigorous, and it is easy to see why trainee novelists might find it hard going.

Comparative philology did not wither, but moved away, far from the turf wars between Lit and Lang, where it has been living quietly ever since under a variety of assumed names. Had Tolkien been working today, he might have found his audience of English scholars and linguists swelled by anthropologists, paleontologists, and geneticists, concerned, as he was, with roots and beginnings, and in charting the history of language by comparative methods.

This equation — between philology and science — would have been hard to make in Tolkien's lifetime. The fault lay partly with evolutionary biology, which was at that time far less scientifically rigorous than comparative philology. Until the early 1970s, evolutionary biologists charting the patterns in the history of life worked in a way more akin to mainstream literary criticism, basing their findings more on subjective notions of plausibility than on objective test.

Ideas about the transition of animals from water to land offer a good example of how evolutionary biology has caught up with the rigor of the nineteenth-century philology in which Tolkien had been schooled. The first amphibians — the distant cousins of modern frogs, toads, and newts — appeared on Earth around 360 million years ago. These primitive creatures bore many resemblances to certain prehistoric fish that had fleshy fins supported by robust bones. From that, it is easy to make the assumption that the amphibians evolved from lobe-finned fish; and from *that*, it is easy to construct a scenario — a 'story' — positing the *causes* of this evolution. For example, one might imagine that fish, living in a drying climate, found themselves in ever-shrinking pools and were driven to evolve legs so that they could scramble over land in search of new bodies of water to colonize.

There seems no doubt that amphibians evolved from lobe-finned fish, but to *impose* such a narrative arc on the process can never be justified, because we have no sure knowledge that natural selection, the driving force of evolution, worked in any such directed way over the course of many millions of years.[1] Scenarios such as the one above are therefore based on subjective appreciation of what is plausible, not on the dispassionate exploration and evaluation of a range of possible scenarios, no matter how outlandish. As it happens, this particular scenario was exploded with the discovery that amphibians evolved legs while still almost wholly aquatic, for reasons we can hardly guess — but certainly not for walking boldly on land.[2]

The works of evolutionary biology thus bore all the hallmarks of the kind of English literature study against which philologists such as Tolkien were fighting their long and ultimately fruitless war of attrition — that is, the works of biology seemed to be more concerned with plot, character, and motivation than the search for origins. To be blunt, scenario-building science is not science at all, its methods owing more to 'Lit' than to Lang.'

The change came slowly. It started in 1950 with the publication of a book by an entomologist working in what was then East Germany. The book's author, Willi Hennig (1913-1976) realized that evolutionary scenarios such as the one above are no help if you are trying (as he was) to unpick the evolutionary relationships of a mass of species of insects, all of which looked very similar. You could, of course, classify your insects based on their overall similarity, but that would tell you nothing about their evolutionary history: that is, the shape of the tree of life, the branching order of the various lines of descent.

Faced with this problem, Hennig invented a logical scheme of classifying things based solely on their evolutionary history. This scheme, which he called 'phylogenetic systematics,' sought to distinguish between those signs of evolution that signify genealogical relationship from those that speak only to similarity or resemblance for other reasons.

Even more rigorously, Hennig's system made a clear distinction between the *pattern* of evolutionary relationship (who was related to whom, and in what degree, in true Hobbit fashion) and the *process* whereby that evolution took place (arguments based on the subjective plausibility of causation, for example, about legs evolved from fins as an adaptive response to habitat changes to which fish were subjected) — concentrating on the pattern as the only topic about which one could ask questions that were legitimately scientific. And what *of* this pattern? At the heart of Hennig's system was Darwin's presumption that the pattern of evolution was like a tree, with a single stem, denoting the single origin of life, and branching through bifurcating boles and branches into the innumerable twigs that represent each one of the species alive today. If Darwin's model of evolution were the true one, Hennig reasoned, it should be possible that the tree of all life, if fully known, could be broken down into nested sets of bifurcations, in which any one species split (by whatever means) to give two descendant species.

Hennig's book was translated into English as *Phylogenetic Systematics* and published in 1966, whereupon the discipline became generally known as 'cladistics' – from a Greek word meaning a branch. Cladistics took another fifteen years or so to make itself felt among scientists. Its spread was due in large part to paleontologists — accustomed, like philologists, to dealing with the fragmentary remains of the past, and for whom the problem of unscientific scenarios of evolution was most acute. Cladistics in one form or another is now the generally accepted philosophy of systematics, that is, the organization of living things into schemes of evolutionary cousinhood.

In time, evolutionary biologists found that the principles of cladistics were more generally applicable than Hennig might have realized. In that it removes all considerations that limit its usage to specific instances (such as the specific problem of the transition of animals from water to land), cladistics can be applied to *any* system, organic or otherwise, in which the members reproduce, evolve and produce a bifurcating pattern of evolution. For example, researchers realized that cladistics was precisely congruent with methods used in a subfield of textual analysis called stemmatics.[3]

Before the invention of printing, the only way to reproduce a manuscript was to copy it by hand. Mistakes invariably crept in: stemmatics is a discipline in which medievalists, faced with a pile of hand-copied manuscripts, seek to work out their copying history by charting the distribution of these mistakes. In the same way that mutations build up in evolution, mistakes tend to accumulate during the copying history of a manuscript, especially as many copies are made from other copies rather than the original. By plotting the occurrence of copying errors (without reference to meaning), a scholar could come up with a diagram summarizing the most likely copying history. This diagram, called a *stemma*, would be, by and large, tree-shaped, and is formally equivalent to a *cladogram*, the tree-shaped diagram drawn up in cladistics to summarize evolutionary change. Stemmatics allows scholars to trace a manuscript through a stream of copies to find the original reading of a text, or something close to it. Stemmatics is now seen as identical with cladistics, and analyses of the history of texts use substantially the same language and methods as are found in evolutionary biology. When stemmatics is discussed in the scientific literature, the term 'cladistics' is often used as a substitute, being more familiar to the scientific as opposed to the medievalist readership.

Sometimes the results are surprising. In a cladistic analysis of 58 fifteenth-century manuscripts of the Prologue to the *Wife of Bath's Tale*, one of Chaucer's *Canterbury Tales*, originally written in the fourteenth century, Adrian C. Barbrook and colleagues identify a corpus of long-neglected manuscripts which are likely to be closest in their reading to the original version, and so could be crucial in elucidating the history of the text. The overall result, however, was that there seems to be *no* satisfactory single reading of the text that can be said to be definitive. It seems as if Chaucer used a collection of working drafts from which he had yet to create a finished version, and copies were subsequently made from each of these variants.[4] This work shows how techniques more usually associated with evolutionary biology can be used to resolve problems in literature, as well as offering an insight into the working methods of an author who lived hundreds of years ago. One wonders what Tolkien would have thought of such powerful methods of analysis, had he lived to see them. If he was a scientist, as he said he was, he might well have approved.

Tolkien's self-affirmation as a scientist seems at variance with the anti-scientific stance he is usually portrayed as having adopted. The same can be said of philology itself. Despite Gandalf's disapproval of Saruman's reductionist tendencies, the comparative philology that occupied Tolkien as a professional academic is very much this kind of exercise — breaking up words to discover how they 'work;' their roots, and their changes in meaning; and how these findings illuminate the histories of ancient peoples.

One of the most important findings in the Germanic, classical philology with which Tolkien was familiar was that many widely dispersed and superficially distinct languages appear to have a family resemblance, hinting at a deep evolutionary history. So it was that a range of languages from Swedish to Sanskrit and Greek to Gujerati came to be united in the 'Indo-European' language family, suggesting the spread of an ancient proto-Indo-European language throughout Europe and Western Asia thousands of years ago. Ancient languages of western Asia known only from scarce inscriptions, such as Hittite and Tocharian, are also believed to be part of this family of tongues.

Other groups of languages were found to share familial resemblances, distinct from those characteristic of Indo-European. Finnish and Hungarian, for example, are the sole European representatives of a language family found in northern Eurasia. Hebrew, Arabic, Aramaic, Assyrian and a host of others belong to the Semitic language family. Most of the world's approximately 7,000 languages can be grouped into one family or another, and ambitious researchers have sought for signs of deeper, supra-familial relationships. Some languages, such as Basque, do not seem to be related to any other known language, and may represent very deep strata of prehistory, the only relic of languages once spoken by Europeans before the arrival of the speakers of Indo-European languages, and perhaps even before the arrival of agriculture.

Can we say anything about that long-vanished time — even *when* it was? There are two long-abiding theories about the origins of the Indo-European languages. Even though the ultimate ancestor of the Indo-European languages no longer exists, we can discover something about it by the traditional, comparative methods that underlie both philology and evolutionary biology, reconstructing vocabularies of the words that might have existed. Words are the spoken avatars of concepts, and the shared roots of words indicate shared experience and culture. Hypothetical ghostwords, formed by comparison of existing words and working backwards through changes in sounds and structure, are conventionally marked with an asterisk, to indicate their provisional nature.

From the shadowy reconstructed vocabulary of Indo-European, scholars have built up a picture of a nomadic people living on some distant grassland, with an economy built around the horse, whose Indo-European cognate is *ekwos*. Archaeology has confirmed that roughly 5,000 years ago, a nomadic people known as the Kurgans lived on the steppes of southern Russia, a people who had a special reverence for horses, and who could have been the original speakers of Indo-European. These horse-lords were thought to have conquered Europe and western Asia long ago, sweeping all before them and implanting their language from Delhi to Donegal.

Another theory has emerged, less dramatic than the Kurgan hypothesis, but backed up by genetics. It concerns the origins of agriculture, and how nomadic peoples settled down and started farming. Agriculture was invented several times in different parts of the world more or less simultaneously around 10,000 years ago.[5] The agriculture that has long been the foundation of settled communities in Europe and the Near East, built on the domestication of such staples as sheep, cattle, wheat and barley, started in a belt of land called the Fertile Crescent, stretching in a great arc from what is now Israel, through Syria and southern Turkey, and down the valleys of the Tigris and Euphrates in central and southern Iraq.

The languages spoken in this region today are not Indo-European. However, the earliest roots of Indo-European, judged from the geographical distribution of Indo-European languages, suggest that it originated in Anatolia, in what is now central Turkey. Archaeological sites in Anatolia include some of the earliest known signs of agricultural settlement, and over the past few decades archaeologists have shown fairly convincingly that agriculture spread north-westwards from Anatolia as a slow but inexorable wave, eventually reaching the west of Ireland and the Outer Hebrides.

This does not mean that farmers scythed their way across Europe in a Kurgan-like conquest. On the contrary, the spread was more peaceful and much slower. Farming supports a greater density of people than foraging, and as people began to settle, the excess of population spread into territory populated only by much more thinly-spread foragers and pastoralists, building new homesteads, taking their farming culture and their language along with them. The spread of farming across Europe took thousands of years. More tellingly, the spread of farming is matched by the distribution of variation in human genes, which shows a clear southeast-to-northwest trend.[6]

Recent work based on certain assumptions about the speed at which sounds in language change (assumptions that are, it has to be said, controversial among linguists), and supported by methods closely similar to cladistics, have produced a family tree of Indo-European languages whose deepest root is more than 9,000 years old — too early for the Kurgans, but consistent with its first speakers having been Anatolian farmers.[7] However, a spurt of linguistic evolution around 5,000 years ago could indicate the activities of the Kurgans: if this is correct, the Kurgans did not invent Indo-European, but were active in spreading it.

Tolkien wrote that the inspiration for all his stories was linguistic. Everything started with the invention of words, their philological dissection and elaboration. These words became embedded in evolving, changing languages, which then required people to speak them. Thus it was that epic tales of entire peoples emerged from the minutiae of words. Tolkien's method of writing stories seems bizarre to any student of conventional literature, for whom names are subservient to character and action, but it describes perfectly the flow of knowledge in modern, biological, evolutionary philology.

Tolkien would have appreciated how work on the tiny details of words found in obscure texts or scratched on ancient monuments — perhaps as the last and only relic of the culture, mythology and even the existence of entire peoples — could, eventually, illuminate the previously hidden histories of peoples and the emergence of our own civilization. Each word is a single leaf, each lovingly painted by a philological Niggle, but which contributes to the vastness of the Tree of Language, and the Tree of Humanity.

3. LINGUISTIC CONVERGENCE

One of the problems that cladistics was designed to address was teasing apart those features of any two creatures that betray a genuine evolutionary kinship, from those that token a resemblance for any other reason, or no reason at all. This problem — called 'convergence' or 'parallelism' — is actually impossible to solve, and yet convergence is believed to be rife. The function of cladograms is not, therefore, to describe what might unequivocally be called the 'true' pattern of evolutionary relationship, but only that pattern that would happen were instances of convergence kept to a minimum.

Philology, too, has its convergences – that is, words in otherwise widely sundered languages that sound and mean the same. Tolkien himself describes one of these in the context of *The Notion Club Papers*, concerning the Old English word *Earendel*, which turns up in the devotional poem *Crist* of Cynewulf. In that context it is thought to mean a ray of light, perhaps in the sense of a herald, an allusion perhaps to the habit of the planet Venus to run before the dawn, as well as to the ministry of John the Baptist that paved the way for Jesus.

In Elvish, however, Eärendil means a 'great mariner,' and one of the characters in *The Notion Club Papers* describes the thrill of recognizing in the words of Cynewulf an echo of an older, vanished language, even if the etymology of the word was different. This draws on Tolkien's own experience as a young man of coming across the line in the *Crist* that reads *Eala Earendel engla beorhtast, ofer middangeard monnum sended*: 'Hail Earendel brightest of angels, sent over Middle-earth to men' (*Biography*).

The characters in *The Notion Club Papers* offer the following real example of linguistic coincidence. The Tamil language of southern India contains a word *popol*, which means 'people' or 'popular assembly.' This sounds, inescapably, like the Latin *populus* and its derivatives (words such as 'popular' and 'population') that mean much the same thing. Does this single example betray a deep, evolutionary connection between Latin and Tamil? The fact that this is an isolated case of resemblance amid a host of differences suggests that it is, in evolutionary terms, a case of convergence. Examination of the cultural context resolves the issue: in Tamil, the word *popol* originally has nothing to do with people or assemblies, but derives from the name for the kind of mat or rug on which the assembled people sat during their convocations.

A more involved linguistic coincidence runs through Tolkien's work, and its persistence suggests that deeper forces are at work apart from chance alone. That coincidence is the repeated presence of words with forms resembling Hebrew. Hebrew is a member of the Semitic family of languages, sharing a history and evolution with modern languages such as Arabic and extinct ones such as Akkadian, an imperial language of Mesopotamia and one of the earliest languages to have been written down. Hebrew in its construction, etymology and phonology (that is, the distribution and usage of sounds) is completely different from the northern languages on which Tolkien based his invented linguistics.[1] The style or aesthetic of High Elven, or Quenya, owed much to the Finnish that Tolkien encountered in the *Kalevala*, whereas the style of Grey-Elven or Sindarin was based on Welsh. On this basis, the 'atmosphere' of Hebrew is all wrong. This hard-edged language betraying a long, sunburned, and above all 'southern' upbringing would have been entirely out of tune in the starlit woods of Elvendom.

Nevertheless, many of Tolkien's words, particularly in Sindarin, adopt plural forms that look startlingly like Hebrew. Hebrew nouns, like French nouns (and, as it happens, Old English nouns), have genders, and these affect word form. Plurals of Hebrew masculine nouns end in the suffix *-im*, for example *chaverim* for '(male) friends' while feminine nouns take the suffix *-ot* (sometimes written *-oth*) for example *succot* ('booths'). Such plurals crop up in Sindarin: examples might include *Rohirrim* (the people of Rohan), *Naugrim* (Dwarves), *Balchoth* (the Easterlings worsted by the Rohirrim at the Battle of the Fields of Celebrant) and *Lossoth* (the 'Snow-Men' of Forochel described in *Rings* A). Tolkien was at pains to point out that such resemblances were entirely coincidental (*Letters* 144). However, elvish writing, particularly in the ancient Fëanorian mode, tends to admit vowels only as diacritical marks: the same is true of Hebrew, and this is also presumably a coincidence.

Even if one accepts this excuse for Elvish, it doesn't wash for Mannish and Dwarvish languages. Tolkien notes that Khuzdul or Khazadian, the secret language of the Dwarves, is very similar in style to Adunaic, the aboriginal language of the Edain, going so far as to suggest that they had a common origin, and making clear (in *The Drowning of Anadûnê*, in *HOME* IX) that their guttural style is very similar to that of Hebrew. In *The Notion Club Papers*, in which the concept of Adunaic first emerged, Tolkien likens the relationship of Elvish (there called 'Avallonian') and Adunaic to that between Latin and Hebrew, at least as regards their aesthetics, and also notes the faintly Semitic flavor of Adunaic.[2]

This explicit connection makes it all the more significant that some plural forms in Adunaic happen to resemble those in Hebrew, even if similarities with Elvish can be said to have been coincidental. For example, the Adunaic name for the people of Numenor is *Adūnāi* or *Adūnāim*. Not only does this look like Hebrew in style, but it fits in with the mythology of Numenor, especially from the perspective of the Men left to languish under the yoke of Sauron in Middle-earth during the Dark Years, in which visits of the lordly but increasingly predatory Numenoreans were mistakenly conflated with the worship of the dark powers. *Adonai* means 'Lord' in Hebrew and is used as a euphemism for the unutterable name of God.

Further links between Adunaic, Khuzdul, and Hebrew come in asides in letters (see *Letters* 156, 211), in which Tolkien likened the Numenoreans to the Jews of ancient Israel in that they were monotheists with only one place of worship. Tolkien also admitted to thinking of the Dwarves as the Jews of Middle-earth in that they were a wandering people, often alien in their present habitations but maintaining their language, which, while they kept it to themselves, colored the accents of the languages they adopted in their host communities (*Letters* 176).

Another connection with Semitic languages in general is made with a philological pun. In linguistic notes to *The Drowning of Anadûnê* (*HOME* IX), Tolkien records that the word for 'wizard' in Adunaic is *zigūr*, which has the very Semitic-sounding plural *zigīrim*. However, Adunaic has a special kind of plural called a 'dual,' referring specifically to a pair of the objects concerned, and the word for two wizards is recorded as *zigūrăt*. It so happens that the word 'ziggurat' is real, and refers to a kind of tower or stepped pyramid in ancient Mesopotamia: the word itself is an Anglicization of the Assyrian *ziqquratu*. Bearing in mind this Semitic allusion, the word seems particularly apt in the context of Tolkien's legendarium, so much so that its use cannot have been a coincidence. Wizards, especially evil ones, live in towers (one immediately thinks of Saruman and Sauron), such that the concept of wizardry might become conflated with their residences: *Zigūrăt* thus becomes a perfectly good Adunaic translation of *The Two Towers*.

Before one gets carried away with all this biblical architecture, Tolkien noted that the ancient Babylonian city of Erech, mentioned in the Bible, has no connection with the place in Gondor of the same name (*Letters* 297). However, he may not have known that the word Erech is itself an Anglicization of the original name of the city, *Uruk*. Therefore the use of the word *uruk* (an Adunaic word) in the sense of 'orc' really does seem to be a genuine coincidence.

From the same source (*The Drowning of Anadûnê*) we find the Adunaic name of Melkor, which is *Meleko*. This is closely similar to the Hebrew word *melekh* both in sound and meaning. In both Adunaic and Hebrew, the word means a king or ruler in the sense of domination. The words *melekh* and *adonai* appear in the liturgical Hebrew formula that prefaces every Hebrew blessing. From this I do not intend to imply that Tolkien thought of Hebrew in any way as some relic of ancient Morgoth-worshipping cults from the Black Years that Christianity had replaced: only that his use of Semitic forms went much further than simple linguistic coincidence.

The question arises, of course, of why Tolkien used Hebrew in this way. Tolkien was sensitive to the sounds and music of words, and gave both Khuzdul and Adunaic a Semitic phonology as a sign both of ancientry and hard usage. Like Hebrew, these languages reflected the rugged grandeur and authority, through the Judeo-Christian liturgical tradition, of their speakers. As a Christian and a philologist, Tolkien might well have been interested in Old Testament Hebrew, and would have been fascinated by such inconsistencies of translation in which, for example, the King James Bible reads the word 'Nephilim' in the legend of Noah (*Genesis* 6: 1-4) as 'giants,' when in Hebrew it means 'fallen ones.'

Such small linguistic problems might well have sparked whole legends in his mind, of races of giants that existed on Earth and which became enmeshed in human mythology, but which died out leaving as signs of their existence only a few confused and half-remembered epigrams. Indeed, one thinks of the similar confusion concerning the Old English word *eoten* in *Beowulf*, which could be construed either as 'giant' or 'Jute' (*Finn and Hengest*).

The resonance of biblical usage is undeniable, and Tolkien would have been more sensitive to it than most. Names for Men and Hobbits of Biblical derivation crop up all over early drafts for *The Lord of the Rings*, and were exorcised only with difficulty. Tolkien found ways round the biblical-sounding 'Sam' and 'Ham' by rendering them not as Hebraic names but, somewhat circuitously, as contractions of plausible Modern English readings of Old English words, so that Sam = Samwise = *samwís* (simple) and Ham = Hamfast = *hamfœst* (home-fast, that is, stay-at-home). But Tobold Hornblower (the celebrated grower of pipe weed) was once Tobias, and Barliman Butterbur (the landlord at *The Prancing Pony*) was Barnabas Butterbur almost up to publication (See *The Appendix of Languages* in *HOME* XII for more discussion of Hebraic names for Men and Hobbits.)

Less prominent, but more pervasive, is Tolkien's use of a biblical style of narrative, especially when recounting a series of momentous events in rapid succession, as if from a remote distance. In *The Lord of the Rings*, this style hardly appears before the Ring is destroyed, but it is pervasive thereafter, from the chapter 'The Field of Cormallen' (Rings VI,4) right up to the end of the book, with the exception of the scenes at Bree and in the Shire, where this 'high' style would seem incongruous. Many passages in the Bible of this nature have verses that start with the word 'and,' in forms such as 'And it came to pass,' and this tendency is especially notable in certain clusters of verses in which great events are summarized rapidly.

For example, just five of the twenty-six verses of Genesis 4 (the story of Cain and Abel) do *not* start with 'and.' Even though some of the familiar examples come from the New Testament (such as the opening of the Gospel of St. Luke), which was originally written in Greek, the style seems to have been a borrowing from earlier, Hebraic tradition. This unusual form of narrative is distinctively biblical, so that its occurrence elsewhere is often designed to stimulate in the mind of the reader the same sense of importance and awe. Tolkien used this device repeatedly. There are two successive paragraphs in *The Steward and the King*, in which Gandalf and Aragorn stand on Mount Mindolluin, surveying Middle-earth, and Gandalf speaks prophetically and somewhat Mosaically to Aragorn about the Dominion of Men, in which there are no fewer than nineteen 'ands,' nine of them at the beginning of sentences or clauses following a semicolon. The very last paragraph of the same chapter, describing the marriage of Aragorn and Arwen, has seven 'ands' despite being just two sentences long. There is a paragraph consisting of just three sentences near the very end of the book, describing Frodo's departure and voyage across the ocean, in which 'and' is used thirteen times. These are just three examples among many.

What is the source of this grammatical peculiarity? The explanation lies with Hebrew, how it uses certain conventions to overcome its severe typographical limitations, and how these conventions may have been systematically misinterpreted when the Bible was translated into English. The Hebrew alphabet has no punctuation marks; neither has it any numbers, nor is any distinction made between 'upper' and 'lower-case' letters. Hebrew letters must often do multiple duty, serving not only as letters but numbers, punctuation marks and grammatical modifiers, their usage depending on the context. For example, in some words that begin with the letter *vav*, the letter is not only pronounced as 'v' but is used to modify the tense of the verb. It is noteworthy that this additional usage of the initial *vav*, as a tense modifier, is found *only* in *biblical* Hebrew, and not in Hebrew found in other contexts. However, the letter *vav* at the beginning of a word may also mean 'and.' In Biblical Hebrew, in which word order is somewhat flexible, and there is no punctuation or capitalization to indicate where one sentence ends and another begins, the letter *vav* is often found as a place-marker to indicate the first word of a sentence, which — because it is the Bible — is often the first word of a verse. From this it can be seen how this typographical convention was translated such that the Bible, and only the Bible, has the oddity of starting many sentences with the word 'and' — the quirks of Hebrew having been lost in translation.

But if Tolkien used Hebraic tradition both explicitly and in his usage of a biblical style, he very nearly fell victim to a linguistic coincidence which, had it been published as such, might have been unfortunate. In a draft (in *HOME* IX) of the chapter of what was to become *The Scouring of the Shire*, reference is made to the rebellious Hobbits capturing ruffians and thrashing them in the 'Tookus.' This word is a contraction of 'Took House,' analogous to the dialectical contraction of 'workhouse' to 'workus.' *Tookus*, however, has a different meaning in Yiddish, the *patois* spoken by Jews from Eastern Europe, a language that is largely German in derivation though it contains elements of Hebrew, and is written in Hebrew script. That meaning is 'ass,' as in 'backside.'

4. THE POWER OF THE NAME

Underneath words of different histories, and languages that follow different rules of grammar, lies a still unanswered question: why do words, as combinations of letters, attract the meanings they do? Why is water called *water* and not, say, *manxome* or *borogove*? This is a question that Tolkien himself was unable to answer (*Letters* 209): however, it gave him license to create words in a primal, almost poetic way, simply because he enjoyed their sounds, and only then worried about how they might be fitted into the growing phonological and grammatical structures of his languages.

If Tolkien was sensitive to the euphony of words, he was also conscious of the power of a word both as a unique descriptor of the thing described, and as a separate entity, subject to its own evolution.

The power of the name resides not just in the object bearing the name, but also in the person in whom is vested the immense responsibility of creating a suitable name. This point is made very clearly in the breach in *The Lord of the Rings*, when the Hobbits try to understand the place of Tom Bombadil in the fictional universe. The nature of Bombadil is a favorite topic of debate among Tolkien fans. Needlessly so, because Tolkien explicitly introduced him as an anachronistic element — Tolkien used his license as an author to introduce Tom simply because he felt like doing so, irrespective of the conventions of Middle-earth (*Letters* 144). The confusion comes when Frodo asks Goldberry who Tom is, and her answer — simply that he 'is' (*Rings* I,7) — seems unhelpful and uninformative.

However, this answer is not as teasing as it seems: Tom is one of a kind, *sui generis*, and so he does not require a name that might distinguish him from anything or anyone else. He, quite literally, 'is,' and that is all that needs be said (*Letters* 153). If Tom had had a twin brother, Tim, this simple reference would indeed be inadequate, and the twins would no doubt have been at pains to show that each was distinct from the other, and had his own jealously guarded and yet exaggerated attributes, real or imagined. But because Tom Bombadil exists alone, no qualification is necessary. By the same token, he can adopt any name he chooses, or which anyone else cares to confer on him (see Elrond's various suggestions in *Rings* II,2). Names, like the Ring, have no power to dominate or constrain Tom Bombadil.

The reverse is seen in a phenomenon, first noted by the psychologist Carl Jung, in which people tend to have names that reflect their profession. So it is that it is not that hard to find electricians called Sparks, nurses named Cutts and fishermen called Herring. Do peoples' names exact a tyrannical influence such that their hapless bearers must follow their dictates? When technical papers on incontinence, authored by a Dr. Splatt and a Dr. Weedon,[1] were drawn to the attention of *New Scientist* magazine, its readers were invited to send in other examples of what became known as 'nominative determinism.'

This Jungian phenomenon illustrates how satisfying it can be when a name is more than a label, but illustrates some property of the thing named. Nominative determinism is amusing because it points up a distinction we usually take for granted. That is, that the name and the thing named are actually *different* things; that the effort of connecting the two is greater than we might imagine; and so it is satisfying when a person has a memorable name that records some distinctive property of the thing named, making it more than an arbitrary combination of sounds. Tolkien was as sensitive to this distinction as anyone: even in the first few pages of *The Hobbit* , Gandalf castigates Bilbo for remembering the name 'Gandalf,' while forgetting that he, the wizard, 'belonged' to it.

There is a branch of science in which correct nomenclature is everything, and on which the whole of natural history is based. That science is taxonomy. The job of taxonomists is to provide names for species of living creatures. In ages past, the lack of any standard nomenclature made it hard for scientists to get the measure of the natural world. When the same creatures were known by host of names in different languages, it was impossible to know whether the same creature was being referred to in each case: as Elrond offers several names for Bombadil, Gandalf offers several names for himself, giving the origin of each. Gandalf is his name only among Men of the North, but he is called 'Incanús' in the South, 'Tharkûn' by the dwarves, 'Olórin' in the ancient West, and so on (*Rings* IV,5). We know that all these names refer to the same person only because Gandalf tells us that this is so, not by some external reference. Were we to meet a southerner who mentioned having met Incanús, for example, we should only discover that we were talking of Gandalf by comparison of his attributes: both Gandalf and Incanús would have a staff, bristling eyebrows, a pointy hat and a silver scarf, suggesting (but not proving) that we were talking of one and the same person. But if Gandalf were known by the same name everywhere, this confusion should never arise, preferably by a name that reflected one or other of his attributes. As an aside, Tolkien got the name Gandalf from the Icelandic *Völuspá* — the same source for all the Dwarf names in *The Hobbit*. The name *Gandalfr*, however, seemed to stand apart, as an argument could be made for its meaning 'Wand-Elf' — in other words, a Wizard, rather than a Dwarf (*Letters* 25).

In the eighteenth century, a Swedish botanist known as Carl von Linné, known as Linnaeus (1707-1778), decided to organize the confusing mass of animal and plant names then in current use by enforcing a standard nomenclature, so that whatever a creature's name in any given language, it would always have a 'proper' or 'formal' name on which everyone could agree. These names would conventionally be derived from Greek or Latin: ancient languages which, because they were extinct, exerted a universal authority while not exacerbating any particular feelings of national pride, and in which all educated people of the time would have been conversant. Linnaeus' standard nomenclature meant that the animal known as *rabbit* in England, *lapin* in France, and *Kaninchen* in German are in fact one and the same species, *Lepus cuniculus*. '*Lepus*' is the Latin word for a rabbit.

Linnaean names are composed of two parts, a 'genus' name (such as *Lepus*) and a 'species' name (such as *cuniculus*). This strict binomial system replaced a rather more chaotic tradition in which an indefinite number of Latin words were used to describe a species.[2] The 'species' would indicate the particular kind of creature; the 'genus' would suggest a larger group to which several different (but related) species might belong. The genus *Panthera*, for example, contains several species, for example *Panthera leo* (the lion) and *Panthera tigris* (tiger). Some genera (where 'genera' is the plural of 'genus') can contain dozens or hundreds of species; others contain a few, or just one.

Ever since the days of Linnaeus, taxonomists have built up a body of rules, rather like English case law, determining how these names are made and used, to ensure that names refer unambiguously to the same things. To reinforce this consistency, whenever some creature new to science is named in the technical, scientific literature, the name is always accompanied by an etymology, that is, an explanation of how the name is made, and the language or languages in which it is written. Very often, a name is constructed to refer to a distinctive attribute of the thing named, and might also honor some person, such as a benefactor.

In addition, the name always refers to a particular specimen of that creature, called the 'type specimen,' held in some recognized museum or collection, so that other scientists can go there and examine it for themselves. Most importantly, a new name is always accompanied by what is called a 'diagnosis.' In just the same way that a medical diagnosis is a list of symptoms characteristic of a particular disease, and that disease only, a taxonomic diagnosis is a list of features one would expect to find in the newly named species, and that species only. For example, a diagnosis of Gandalf would be that he is always characterized by a silver scarf, pointy hat, bristling eyebrows and staff: these features, being found in Gandalf alone, set him apart from other wizards (Saruman has a staff but not a pointy hat, for example) and from other creatures more generally.

This need for standard nomenclature has only become more acute since Linnaeus's time. Linnaeus named a few thousand largely familiar creatures, mainly using the same names that any Roman would have used for the creature concerned. *Equus*, *Ovis*, *Canis* and *Felis* are simply Latin for, respectively, horse, sheep, dog and cat. Today, however, more than a million creatures have been formally 'named,' most of which would be unfamiliar to people in general. It may come as a surprise that the vast majority of animals that exist on the Earth are insects, so there are hundreds of thousands of names for insects, in particular for beetles, and legions of mote-sized wasps.

As the roster of named creatures has expanded, so has the need for new names that have not already been used. One of the main tasks of the International Commission of Zoological Nomenclature — the United Nations of zoological taxonomy — is to assist scientists with problems of nomenclature.[3] Botanists and microbiologists subscribe to their own codes, tailored to their own needs and the peculiarities of their subjects. The need for new names has meant that taxonomists have had to expand well beyond the classical in search of new, unique names.

Names are no longer restricted to Latin and Greek alone, and can derive from any human language. Many names are, in fact, jokes. When Dr. Jennifer A. Clack was describing a new species of fossil amphibian that lived in a fetid swamp, she called it *Eucritta melanolimnetes*, a name broken down as *Eu* (Greek, 'true'); *critta* (American vernacular, 'critter,' meaning 'creature'); *melano* (Greek, 'black') and *limnetes* (Greek, pertaining to lakes) — and there you have it — the Creature From The Black Lagoon.[4] On another occasion, paleontologist Dr. Scott Sampson and his colleagues discovered the fossil skeleton of a fierce-looking dinosaur in Madagascar while listening to the music of Dire Straits — which, they said, always brought them luck while on the fossil trail. They named their new find *Masiakasaurus knopfleri*, whose name breaks down as *masiaka* (a Malagasy word meaning vicious), *sauros* (Greek, 'lizard,' a common element in dinosaur names, even though dinosaurs and lizards are not closely related) and *knopfleri* (in honor of Mark Knopfler, the lead singer and guitarist of Dire Straits.[5])

Many scientists have sought and found nomenclatural inspiration in works of literature, folklore and popular culture.[6] There are ichthyosaurs — giant, extinct sea reptiles — called *Grendelius* (after Beowulf's adversary) and *Excalibosaurus* (after King Arthur's sword); a shark called *Iago* after Shakespeare's villain; and an anglerfish called *Puck*, after the Bard's mischievous sprite. Sophocles' tragic hero is particularly well commemorated, in a worm (*Oedipodrilus oedipus*); a monkey (*Saguinus oedipus*); and a salamander (perhaps inevitably, *Oedipus rex*). Swift's *Gulliver's Travels* has generated a respectable haul of names, including *Balnibarbi* (a trilobite); *Dryadella lilliputiana* (an orchid); and *Holorusia brobdingnagia*: a nine-inch-long crane fly, possibly the largest crane fly ever discovered, and appropriately named after Swift's race of giants.

Lepidopterists are particularly fond of naming butterflies of the family Lycaenidae after the works and characters of Nabokov, partly because Nabokov himself was an authority on this group of butterflies. Hence *Paralycaeides hazelea, Madeleinea lolita, Pseudolucia charlotte, Pseudolucia clarea, Pseudolucia humbert* and *Pseudolucia hazeorum*. There is a spider named *Pimoa cthulhu*, after H.P. Lovecraft's demon; a tiny mite called *Darthvaderum*, and a wasp called *Polemistes chewbacca*, not entirely unrelated to *Polemistes vaderi*. Fans of Douglas Adams' *Hitch-Hiker's Guide to the Galaxy* will be amused by the fish *Bidenichthys beeblebroxi* and the moth *Erechthias beeblebroxi*, creatures with false 'head' patterns, named after Adams' bicephalous hero Zaphod Beeblebrox. J. K. Rowling's Harry Potter stories have made it into the scientific canon, with *Macrocarpaea apparata* — a twelve-foot-high plant that suddenly appeared before botanists exploring a misty hillside in Ecuador, as suddenly as if it had 'apparated,' like one of Rowling's student wizards, out of nowhere.

Given Tolkien's passion for nomenclature, his coinage, over decades, of enormous numbers of euphonious names — not to mention scientists' fondness for Tolkien — it is perhaps inevitable that Tolkien has been accorded formal taxonomic commemoration like no other author. The paleontologist Simon Conway Morris discovered an extinct marine worm and called it *Ancalagon*, after Tolkien's altogether more spectacular, fire-breathing variety.[7] It was Conway Morris who also named a bizarre, unclassifiable fossil *Hallucigenia*, after the dreamlike state induced by studying it[8] — a creature used by science fiction author Greg Bear as the model for an alien race, the Jarts, in his novel *Eternity*.

There is a shark called *Gollum*, a wasp named *Gwaihiria*, and a pair of weevils named *Macrostyphlus gandalf* and *Macrostyphlus frodo*. There are no fewer than thirteen tiny moths of the genus *Elachista* that are named after Elves, partly because they are hard to see, and have spread from Europe (Middle-earth) to the Americas (the West). There is a beetle called *Pericompsus bilbo*, because it is short and fat with hairy feet.

But the prize for Tolkien-related obsession in taxonomy must go to paleontologist Leigh Van Valen, who in a single paper named virtually an entire fauna of fossil mammals after Tolkien characters.[9] So, roll up *Thangorodrim thalion, Protungulatum gorgun, Oxyprimus galadrielae, Deltatheridium durini, Arctocyonides mumak, Platymastus palantir, Platymastus mellon, Mimotricentes mirielae, Desmatoclaenus mearae, Litaletes ondolinde, Bomburia priscus, Protoselene bombadili, Tinuviel eurydice, Fimbrethil ambaronae, Mimatuta morgoth, Mimatuta minuial, Earendil undomiel, Anisonchus athelas, Anisonchus eowynae, Anisonchus (Mithrandir) oligistus, Niphredil radagasti* and *Ankalagon saurognathus* (renamed from *Ancalagon* after Van Valen discovered that Conway Morris had got in first, the previous year, with his extinct worm, which therefore had taxonomic priority.)

The progress of science will no doubt ensure that some of these creatures — described, in many cases, on the basis of very fragmentary remains — will turn out to be as mythical as their namesakes. But just one fragment is enough to make a name, which is valid only so long as you can point to a real live (or dead) specimen on which the name can be based.

However, this has not stopped people creating fictional names, either seriously or in jest. The late British naturalist Peter Scott was notorious for applying the name *Nessiteras rhombopteryx* to the Loch Ness Monster, based on a photograph of what might have been a rhombus-shaped fin (hence '*rhombopteryx*').[10] By the rules of taxonomy, this name is invalid because Nessie has never turned up to claim her name in any more explicit way. (This kind of name, that is, a name without anything to attach to it, is useless, and known in the trade as a *nomen nudum*: this is the reason why Gandalf is so disparaging of Bilbo's careless use of his name in *The Hobbit*.) If that were not so, names of fictional creatures would be as valid as the names of real ones, and we would have no way of deciding, when presented with mongooses and momewraths, which one is the more 'real.'

Of course, the majority of 'real' creatures that receive formal names might as well be as mythical as Elves and dragons, given that most people are unlikely ever to meet them. When, for example, did you go swimming with *Iago* or *Gollum*? Tolkien was sensitive to this, particularly for extinct creatures, which are, when all is said and done, no more broken fragments of bone recreated with a dash of human imagination and artifice into the seemingly living, breathing dinosaurs of popular entertainment (*Letters* 211). This is why having type specimens to represent a named organism is crucially important, because a creature without one might as well be as mythical, in formal terms, as an Ent. But because most people think of such things as dinosaurs as real animals, and because only specialists will appreciate the importance of type specimens, the dinosaurs of popular culture really are the same as dragons.

I have exploited this nicely on two occasions: once to make a serious point, the other as an April Fool joke. I used Tolkienian names on both occasions, and both times people were confused by the pervading mythology of names.

The first was in a book I wrote called *A Field Guide to Dinosaurs*, which was splendidly illustrated by my friend the artist Luis V. Rey (who, incidentally, painted the cover for this book.) In our book, we wanted to convey to the reader of 'being there,' in the Mesozoic Era, among dinosaurs as living animals. We explicitly said that the book was fictional, an entertainment: but like all good fiction, founded on a core of well-researched fact. We also noted that the record of Mesozoic life, as judged by the fossils preserved from that long-vanished age, was minuscule. For all his wide-screen dominance, there are rather few specimens of *Tyrannosaurus rex*, and most species that ever existed have left absolutely no fossils at all. These include such things as parasites, so to this end I invented, among other things, *Praefasciola brachiosaurae*, a ten-foot liver fluke that inhabited the insides of giant sauropods, and a fly, *Luisreya ginsbergi*, that carried the embryos of the fluke between dinosaurs. I also created dinosaurs offstage: while discussing *Tyrannosaurus rex*, I mentioned its (fictional) cousin that hunted dinosaurs on the North Slope of Alaska — *Tyrannosaurus helcaraxae*.

The serious point was that the world of the past was far richer and stranger than we might imagine from the few specimens that paleontologists have discovered, and some impression of this vast, lost diversity had to be given if readers were to feel that they really were in the Mesozoic, and not in a museum. And yet only a single, real specimen need divide the commonplace (*T. rex*) from the entirely mythological (*T. helcaraxae*). The dividing line is so fine that the use of fiction in *The Field Guide to Dinosaurs*, even though explicitly stated, made some commentators uneasy.

My second invention was published on the website of the science journal *Nature* on April 1, 1998. It was a report of the discovery of a fairly complete specimen of a carnivorous dinosaur, whose unusual features included scorch marks on the internal surfaces of the neck vertebrae, and a pair of spindly, wing-like arms growing from the shoulders, in addition to the shorter, regular pair of forelimbs. I expect you can see where this is headed: the report was based on a technical paper in the journal *Plains Paleontologist*, authored by one Randy Sepulchrave of the University of Southern North Dakota at Hoople, and naming the strange new creature *Smaugia volans*. This story is so outrageously full of fictional references that anyone should have spotted it. Apart from the Tolkienian *Smaugia*, there is no such serial as *Plains Paleontologist* (although a magazine called *Plains Anthropologist* really does exist, and is a reputable journal of record); 'Sepulchrave' is the family name of the Earls of Groan in Mervyn Peake's *Gormenghast* books; and the University of Southern North Dakota at Hoople is the fictional home of the musicologist Prof. Peter Schickele, who with his students is devoted to the rediscovery and performance of the works of P.D.Q. Bach, the last and most definitely the least of the sons of the great Johann Sebastian.

But still people fell for it. A national newspaper in Spain ran the story and was enraged to learn that they had been spoofed. (A paper in Portugal got the joke at once, however, and asked *Nature* if it had any more.) A correspondent wrote earnestly to a dinosaur newsgroup asking for more details about the habits of *Smaugia volans*. Such is the power of the name: it can delude and deceive, as well as entertain and inform. Which is why taxonomists — and Tolkien — have always been very careful in their creation and use.

5. HOLES IN THE GROUND

In common with other fantasy authors of his era, Tolkien was fascinated by the submerged and the subterranean. The list of underground locales in Tolkien is a long one, including homely Bag End (*The Hobbit* even starts with a 'hole in the ground'); the grandeur of Moria and the Kingdom under the Mountain; the horror of Shelob's Lair and the Paths of the Dead; and the squalor of Gollum's Cave and the other innumerable rat-holes in the Misty Mountains. It is notable that virtually all the substantial centers of population in Beleriand (the large region on the western shores of Middle-earth wherein was set most of the action in *The Silmarillion*) are underground, whether they are the subterranean throne of Morgoth, the Dwarvish cities of Nogrod and Belegost, or the Elvish palaces of Menegroth and Nargothrond. The Elvenking's Halls in *The Hobbit* are a direct and detailed copy of these earlier strongholds, right down to their situation in sheer cliffs accessible only by a bridge over a ravine. Even Gondolin, itself in the open air, can only be reached by a tunnel.

This fascination echoes an underground tradition also seen in stories as diverse as Jules Verne's *Journey to the Centre of the Earth* (published in England in 1871), H.G. Wells' *The Time Machine* (1898) and *Land Under England* (1935) by Joseph O'Neill (1878-1953), a novel that Tolkien specifically mentions as having read (*Letters* 23).

It also resonates with Arthurian legend — compare the cavern in which King Arthur's knights sleep with the Barrows on the Barrow-Downs, not to mention the retelling of Arthurian themes in the *Weirdstone of Brisingamen* (1960), by Tolkien devotee Alan Garner — a novel set in caverns beneath the real-life Alderley Edge in Cheshire. One can only speculate on the reasons for this tendency.

Whatever its source, this fixation with subterranean landscapes also encompasses submerged ones. As well as Numenor, drowned at the end of the Second Age, Tolkien finished the First Age by inundating Beleriand. Tolkien was sensitive to the idea of inundation through geological change: Middle-earth is meant to be the northwest of the Old World in a geologically distinct epoch from our own, and he once wrote that the name 'Gondwanaland' was the closest approach made by geology to poetry (*Letters* 324). This shows that Tolkien must have at least heard of the theory of continental drift proposed in 1915 by Alfred Wegener (1880-1930), outside of which 'Gondwanaland' — as a name for the once-united southern continents – has no meaning. Wegener's theory did not, however, receive general acceptance until the 1960s, so geologists were restricted to catastrophic inundations of the kind that appear in Tolkien's novels which, to my mind, owe more to the apocalyptic paintings of John Martin (1789-1854) than to science.

A peculiar nod to Tolkien's obsession with lost worlds underground or beneath the sea appears in a scientific paper, "Marine Geology of the Rockall Plateau and Trough," by David G. Roberts.[1] This paper concerns the geography of the Rockall Plateau, a submerged fragment of continental crust, isolated when Greenland, North America, and Europe split up when the North Atlantic opened up between about 75 and 60 million years ago. Slightly larger than the island of Ireland, at least down to the 1000-fathom line (a fathom is equivalent to about 1.8 meters), the Rockall Plateau is bounded to the south and west by steep slopes down to the mid-Atlantic, and to the east by the 3000-meter-deep Rockall Trough that separates the Plateau from Britain. The northern boundary is less clear — the Plateau grades into another feature of seafloor topography endearingly called the Lousy Seamount.

The only part of the Plateau that pokes its head above the sea these days is Rockall, a spike of ancient granite just 30m by 25m and standing just 19m proud of the ocean, visited only by depressive seabirds, SAS commandoes, and Greenpeace activists. In the still-unresolved debate about the political status of Rockall, William Ross, the Member of Parliament for Kilmarnock in Scotland, told the House of Commons in 1971 that more people had visited the Moon than had set foot on Rockall. There is evidence that Rockall was once rather larger, and that there were other islands like it, during low sea-stands in the past few thousand years, but not enough to make any appreciable landmass. This has not prevented fantasy writers and political satirists inventing such things, however,[2] and the very presence of Rockall and other remote islets in the North Atlantic could have inspired and informed *Imram*, Tolkien's poem about the legendary Celtic voyager St. Brendan (see *HOME* V), not to mention other western islands more central to the legendarium, such as Tol Eressëa, the Enchanted Isles, and Numenor itself: Rockall might be the peak of Meneltarma, in legend the highest peak of Numenor and the only part not submerged, from which far-sighted men might see the Blessed Realm.

If we imagine that the Rockall Plateau rose by 1000 fathoms, so that the 1000-fathom line of the Rockall Plateau became a seashore, the Plateau would become an underwater 'island,' built of a series of highlands — the Rockall, Hatton and George Bligh Banks — arranged around the central Hatton-Rockall Basin. In his paper, Roberts reports several deeper and hitherto unknown banks to the southwest of the Plateau, naming them the Lorien, Fangorn and Edoras banks. The disposition of the various features, however, bears no relation to the relative placement of the analogous features in Middle-earth. Further to the southwest, off the southwestern 'coast' of the Plateau, are isolated submerged mountains known as the Rohan and Gondor seamounts, which are mentioned in the text even more casually than the other features.

A fold-out map contains further Tolkien-themed features not mentioned in the text at all: an Eriador seamount (close to the Rohan seamount); Gandalf's Spur, an escarpment to the east of Lorien bank, and Helm's Deep, a submarine canyon between Edoras and Fangorn banks. The topography of Helm's Deep — a canyon trending and narrowing northwards as it delves into higher ground — is the only Tolkien-themed feature that bears any relationship to the cognate feature in Middle-earth.

Why was Roberts inspired to name so many features of the Rockall Plateau after names and places in *The Lord of the Rings*? The paper itself gives no clue. Was it to mark a shared fascination, with Tolkien, with drowned continents, particularly in the western Ocean that had fascinated the *Beowulf* poet? The reason, it turns out, was much simpler – that Roberts enjoyed *The Lord of the Rings*,[3] and wished to mark his pleasure in the form of names for a still largely unexplored part of the submerged world, whose features he had discerned largely single-handed — as if the Rockall Plateau was his very own Middle-earth.

It has occurred to me that the line of the continental shelf off north-west Europe looks startlingly like the northwest of Middle-earth, with The Bay of Biscay matching the Bay of Belfalas. In which case, Roberts might have been moved by Tolkien's view that the northwest of his Middle-earth is meant to stand for our own parts of the world, albeit at some prehistoric age so remote that the shapes of the lands would have changed in the interim. As geologists now recognize, the relative positions of land and sea have changed markedly over the ages. It is remarkable that Tolkien built geological change so substantially into his legendarium — the movements of islands across the sea, not to mention the raising and destruction of mountains and the ruin of Beleriand in *The Silmarillion*; the destruction of Numenor and the 'breaking' of the world at the end of the Second Age — not just as nods to the style of catastrophe as found in Biblical and yet more ancient legends, but as a testament to the very latest movements in geological thought.

6. INVENTING THE ORCS

The origin of the Orcs is arguably the most persistent, difficult, and ambiguous problem of Middle-earth. The war against Morgoth — and later, Sauron — is only made more heroic and desperate when fought against overwhelming odds. But the problem remains: where *did* the Dark Lord get his armies? Were soldier Orcs created from inanimate matter as so graphically depicted for Saruman's Uruks in Peter Jackson's film adaptations of *The Lord of the Rings*? Or were they, perhaps, bred from some other creature that already existed? But if they were created first, might they *then* have been allowed to run wild and breed, *au naturel*? Were *some* Orcs bred, and *others* created? If at least some *were* bred from pre-existing species, which ones? Elves? Men? Both? Or something else? And *if* they bred, *how* did they breed? Did they have sex? Lay eggs? Clone themselves? Or a combination of these? Tolkien wrote about Orcs for more than half a century, revising his ideas and forever changing his mind to such an extent that any one of the hypotheses I've just mooted, or even several of them together, could be made consistent with at least some of the textual evidence.

As with so many of Tolkien's words and concepts, the word 'orc' has very deep linguistic roots. According to the Oxford English Dictionary (OED), it is related to the Latin *orcus*, whence comes the word *ogre*, a monster usually depicted as a ferocious giant in human shape, and now usually applied to a person of overbearing manner or characteristics, such as Lady Bracknell in Oscar Wilde's *The Importance of Being Earnest*: a definition in which the division between Man and Monster can be hard to draw: "I don't really know what a Gorgon is like," says Ernest Worthing, but I am quite sure that Lady Bracknell is one. In any case, she is a monster, without being a myth, which is rather unfair.

Tolkien, however, was probably more directly influenced by *Beowulf* which in one place[1] runs

eotenas ond ylfe ond orcneas
swylce gigantas, þa wið gode wunnon

which may be translated as "ettins and Elves and evil spirits/ as well as the giants that warred with God" from which Orcs (*orcneas*) are just one of a number of ill-defined mythological creatures, lumped in with *eotenas* (whence 'etten' and 'ent,' words for 'giant,' both of which Tolkien exploited.) Benjamin Thorpe, in his 1855 translation of Beowulf,[2] rendered *orcneas* as *orken*, sometimes taken to be the singular of *orcneas*. In his later, unpublished writings, Tolkien used the spelling *ork* rather than orc, because it was etymologically more correct; because it would be more consistent with the adjective derived from it ('orkish' rather than the awkward 'orcish' that might be pronounced 'or-sish'); and also because it distinguished the word from a meaning of the word 'orc' in more current usage — that is, a killer whale (*Orcinus orca*).

An Old English clue to the meaning of the word 'orc' comes in the phrase *orcþyrs oððe heldeofol*: 'orc-giant or hell-devil': orcs are clearly unpleasant, even demonic creatures, but their precise nature remains scarcely more defined than in the phrase about *eotenas* and *ylfe*. The OED cites references to horrible, ogre-like *orques*, *orkes* and Orcs from 1598 onwards, right up to Charles Kingsley in 1865, writing in his novel *Hereward The Wake* about

things unspeakable — dragons, giants, rocs, orcs, witch-whales, griffins, chimeras, satyrs, enchanters, Paynims, Saracen Emirs and Sultans, Kaisers of Constantinople, Kaisers of Ind and of Cathay, and beyond them again of lands as yet unknown.

This passage looks like a retread of the Beowulfian *eotenas*, *ylfe*, and *orcneas* in that it is a list of miscellaneous and largely interchangeable beasts united solely on the basis of their unfamiliarity and, no doubt, their hostility towards our parochial selves. The OED entry for 'orc' culminates with citations from *The Hobbit* and *The Lord of the Rings* and defines Tolkien's use of the word as one of a warlike people in whom are combined human and ogre-like characteristics — a definition that could accommodate Lady Bracknell's barbed gentility and all the hounds of Hades with equal facility. If Tolkien left the origin and nature of Orcs nebulous, this is no more than how he found them.[3]

The first detailed published descriptions of Orcs (in *The Hobbit*) say nothing about their origins, but much is implied about their variety. The Great Goblin, the leader of the tribe of Orcs met by the travelers in the Misty Mountains, is described as being distinctly bigger than its fellows. In addition, passing reference is made to a number of different *kinds* of Orc. The term 'goblin' can be seen as a generic word for an Orc in what is, after all, a tale for children, but it is sometimes used to mean, more specifically, a member of one of the smaller kinds of Orc: the creatures properly termed 'orcs' are larger than mere 'goblins,' and passing reference is made to still larger varieties, such as 'hobgoblins.' From *The Hobbit*, at least, our picture of Orcs is of an indeterminate number of distinct and more or less isolated colonies or tribes of creatures that differ physically — and, one assumes, genetically — from one another.

Such physical differences are emphasized in *The Lord of the Rings*: the clearest example occurs in Mordor, when Frodo and Sam are being tracked by two soldier Orcs, one a large warrior, the other a tracker — a member of a smaller species with an enhanced sense of smell (*Rings* VI,2). The larger Orc continually makes racist remarks at the expense of the tracker, which eventually results in murder. This scene reminds me of a similarly militaristic passage in Salman Rushdie's *Midnight's Children* in which the hero, who has an enhanced sense of smell, is used as a tracker by the army and suffers similar torments from his superiors.

As shown very clearly in *The Lord Of The Rings*, such physical differences are underscored by linguistic ones, and it is repeatedly made clear that Orcs from different tribes cannot understand one another's native speech, and so they are constrained to use the Common Speech when in company, albeit in crude form (see for example *Rings* III,3). Tolkien attributes this debasement to the unloveliness of the Orcs themselves, although — were one to be charitable — it can be assumed that the Common Speech used by Orcs was a kind of language that linguists refer to as a 'pidgin.' That is, a jargon constructed to facilitate the communication between people of different origins, largely for reasons of trade or business.

The grammar and syntax of pidgins are, at least to start with, typically much less sophisticated than in 'proper' languages, reflecting their origins as simple means of communication invented by adults of different origins trying to find some common ground, rather than as birth languages with a long and rich history of literature and usage. A good example of a pidgin is 'Russenorsk,' a mixture of Russian and Norwegian developed by arctic fishermen so they could communicate about matters of common interest, such as fishing and the weather. Russenorsk died out around 1917, and was never sophisticated enough for discussions of, say, poetry or philosophy. Other pidgins, such as those found in New Guinea or Hawaii, have evolved further, having acquired a literature and the necessary standardization of grammar. As Steven Pinker describes in *The Language Instinct*, when spoken as a birth language by the children of pidgin-speaking people of different ethnic origins, pidgins acquire the grammar and syntax of a fully-fledged language and become *creoles*.

In its triumph of resourcefulness over lack of vocabulary, argots such as the pidgin spoken in New Guinea (a language known as *tok pisin* — a representation of a Chinese pronunciation of the English 'talk business') have an undeniable charm, as evidenced by the official pidgin designation *Feller Bilong Mrs. Queen* for Prince Philip, the husband of Queen Elizabeth II. It would, however, be a mistake to assume that this or any other pidgin does not have rules, simply because, in their origins, they sound like they were made up literally as speakers went along. Nevertheless, one should not assume that Orcs of various kinds, meeting up occasionally for particular missions, such as the Orcs from Isengard, Mordor, and Moria in *The Two Towers*, would have spoken any more than the most rudimentary kind of pidgin, based largely on words from the Common Speech mixed with elements of Black Speech, indigenous orkish dialects, and even Elvish: Tolkien derives the orkish word *tark* from a Quenya word *tarkil*, used in the Common Speech to refer to a person of Numenorean descent (*Rings* F): *tark* is a good example of the kind of word that might have occurred in a pidgin used by Orcs of different origins when they met.

Perhaps the best analogue from the human world for the meeting of physically, linguistically and mutually suspicious Orcs is the situation in which the many tribes of the New Guinea highlands find themselves. As Jared Diamond describes in *The Rise and Fall of the Third Chimpanzee*, these tribes, separated from one another by forested and mountainous terrain, have diverged physically and linguistically to the extent that each has its own language, as different from the one spoken in the next valley as French is from Finnish. And, when members of different tribes meet, the occasion is sometimes hostile, as is also invariably the case when Orcs of different kinds are forced to work together for any length of time. This hostility is crucially important for the plot of *The Lord of the Rings*: Merry and Pippin manage to escape to Fangorn for their momentous meeting with Treebeard only after taking advantage of confusion generated by the mutual hostility of the various tribes of Orcs taking them to Isengard. In the same way, Sam is only able to rescue Frodo from the Tower of Cirith Ungol because the Orcs from Minas Morgul and Barad-Dur have slaughtered one another, leaving none alive to guard the prisoner.

The enormous variety of Orcs — which, as it turns out, is crucial to the story — can be seen as a consequence of the smallness and isolation of populations evolving in their own particular ways to suit local conditions, their isolation enhanced by mutual antipathy and incomprehension. Evolutionary theory tells us that evolution happens faster, and has more idiosyncratic results, when populations are small and isolated, so Tolkien's portrait of the Orcs as a collection of very diverse kindreds is biologically very accurate. Except, that is, for one thing — sex.

In general, the rapid evolution and diversification of species is strongly enhanced by sex, the effect of which is to spread new and advantageous mutations rapidly throughout a population. However, there is little evidence for sex of any kind in Orcs. Apart from references to their breeding or manufacture by others (which I shall discuss later), there are only four references to orkish reproduction in Tolkien's published works.

One, in *The Hobbit*, mentions the goblin 'imp' that Gollum had killed not long before meeting Bilbo. Two more, in *The Lord of the Rings*, refer to Orcs having 'spawned,' as if they were frogs or mushrooms (*Rings* II,5; VI,1). A fourth describes an Orc chieftain called Azog, and his son, Bolg, involved in the War of Dwarves and Orcs towards the end of the Third Age (*Rings* A). However, the imp might have been a small, adult goblin rather than a goblin child, and references to spawning could be seen simply as derogatory and without specific reference to reproduction, equivalent to the wish that one's enemies should crawl back under the stones whence they came, as if they were not human beings but some lower form of life.

The discussion of Azog and Bolg seems to be evidence of a mode of orkish reproduction no different from that seen in other creatures, but compared with the detailed discussion of the family relationships of other speaking peoples, it is sketchy indeed: in contrast to Elves, Men, Dwarves, Hobbits and even Ents, no specific mention is made of female Orcs. These references, therefore, need say nothing to demand that orkish reproduction is confined to the traditional boy-meets-girl mode. And without any further confirmation, the stated relationship between Azog and Bolg as father and son need not be taken at face value, any more than we need believe that Frodo is literally Bilbo's 'nephew,' as described, when the two Hobbits were, in fact, cousins of a more remote degree.

However, this is not to say that Orcs did not reproduce. After all, Tolkien's picture of Middle-earth was, by his own admission, incomplete, and deliberately so — and Tolkien was particularly reticent about sex. Some have gone so far as to say (somewhat archly) that there is, in fact, *no* sex in Middle-earth — a circumstance which Brian Aldiss and David Wingrove note in *Trillion Year Spree*, their critical history of science fiction, is very strange, given the large amounts of sword-fodder necessary to make the story work. However, the fact that Tolkien doesn't specifically note the existence of something in Middle-earth does not automatically mean that the unmentioned thing did not or could not have existed: all that sword-fodder had to be produced *somehow*.

On the other hand, the lack of any unambiguous reference to female Orcs could be significant given that Tolkien, despite his reticence, is otherwise quite clear that there are, explicitly, *two* sexes, male and female, in other 'speaking peoples.' If sex is referred to very rarely in any direct way, its evidence is seen everywhere in the extensive genealogies in *The Lord of the Rings* and *The Silmarillion* for Hobbits, Dwarves, Elves and Men. Even Ents (for whom there are trees, if not family trees) once had 'Entwives' and 'Entings.' Apart from the seemingly anomalous and exceptional reference to Azog and Bolg, no such statement is ever made concerning the products of the forces of darkness. Orcs — along with trolls, dragons, balrogs, and so on — behave as if they were sexually neutral. So much so, that even though the members of the armies of darkness are always referred to as if they were male, this could reflect little more than the conventional use of English in Tolkien's time, in which a person was always referred to as male, if no specific attribution of gender were otherwise made or implied.

It would be very simple to attach a moral meaning to such neutrality. That is, the forces of darkness are so devoted to conquest and domination (they are literally hell-bent on it), that all resources that might have been spent on sexual reproduction have been subverted by warfare. In place of organic, even romantic sexual activity (after all, even dwarves have love-lives), there is starkly efficient and soulless manufacture. For Tolkien, the use of industry to mass-produce not just weapons and engines but sentient beings, especially for use as sword-fodder, would represent the most depraved perversion of power that there could possibly be.

Such emasculation carries echoes of the mechanization ushered in by World War I, in which Tolkien had had immediate experience; in which dashing horsemen were replaced by soulless and much more destructive tanks, with all that implies for concepts such as individual bravery and the acquisition of honor thereby. The contrast between the efficient Orcs, played with all the calculation of a chess master, and the recklessly heroic Rohirrim, riding joyfully to almost certain death, becomes all the more stark if seen through the lenses of a war veteran exposed at first hand to the introduction of mechanized artillery and the consequences of its use. In which case, the sexlessness of Orcs could, in its own way, make *The Lord of the Rings* as stark a comment on the emasculating effects of industrialization as Dickens' *Hard Times* (1854) or, more pertinently, Aldous Huxley's *Brave New World* (1932) in which sex has gone to the other extreme, in that it has become so pervasive as to lose all meaning. That war veterans such as Tolkien, Mervyn Peake (the *Gormenghast* trilogy), Olaf Stapledon (*Star Maker, Last and First Men*), and Kurt Vonnegut (*Slaughterhouse Five*) pursued fantasy as the only literary mode capable of expressing their extreme experiences within the constraints of literature is a point amply made by Tom Shippey in *J.R.R. Tolkien: Author of the Century*.

If this view — that the armies of darkness were emasculated neuters — seems extreme, then consider that the proof is shown in the breach, with the aggressively female Ungoliant, and her descendant, Shelob. As malevolent spirits in spider form, these creatures are as dark as they get, but Tolkien is always keen to emphasize their independence from the organized war machines of Morgoth and Sauron. *The Silmarillion* shows that Morgoth and Ungoliant had an uneasy truce at best, and in the end, Morgoth was overwhelmed by the spider, and had to be rescued by his balrogs. In *The Lord of the Rings*, Sauron refers to Shelob as his pet cat,[4] even though everyone, including Sauron, knows that Shelob is a law unto herself. The spiders are female, and like all the most memorable vamps in literature, from Cleopatra to the absent heroine of Daphne du Maurier's *Rebecca* (1938), they tie men in knots but are tied to no one, and with this liberty goes their sexuality.

But to return to Orcs: industrial manufacture of the kind I have discussed, that would have reduced all Orcs to neutrality, need not have applied to the wild (or at least feral) Orcs living in isolated tribes in the Misty Mountains, who appear to have some biological (as opposed to technological) means of population increase. As far as Tolkien is concerned, just about anything is possible. It is, of course possible that there were indeed separate sexes of Orc, and that they produced infants in the same way as other creatures: the whole business of Azog and Bolg would, at first glance, support this.

But other options are possible, even likely, and would not in any case preclude a direct lineal relationship between Azog and Bolg. It could have been that Orcs laid eggs, or that Orcs had a population structure similar to that of social insects such as bees, wasps, or ants. That is, individuals in a colony would belong to physically distinct castes, one of which would be devoted to reproduction. Rarely seen 'queen' Orcs, much larger than the others, would produce all the other Orcs in the colony, inseminated by a small band of 'drone' Orcs, with the whole colony served by small, servile 'worker' Orcs and guarded by large, heavily armed 'soldier' Orcs. There is nothing in Tolkien that contradicts this scenario. Indeed, the Orc colony described in *The Hobbit* is very like an anthill reacting to a disturbance (especially when the notably large Great Goblin has been killed). Much the same can be said for the Orcs of Moria in *The Lord of the Rings*. And when we see Orcs away from home, their physical distinctiveness, mutual loathing, and style of military organization suggests roving bands of soldier ants battling for supremacy on the forest floor, killing or enslaving anything else they meet.

To pursue this idea even further, the Orcs could have reproduced without any kind of sex at all, by cloning themselves. It so happens that the females of many kinds of animal are able to engender new individuals without any male input whatsoever, a process that is called 'parthenogenesis' (literally, 'virgin birth'). In 1677, Antoni van Leeuwenhoek, a pioneer of the then new-fangled microscope, reported that female aphids (that is, the greenfly or plant lice that infest your garden plants) could reproduce without males. This discovery was followed up in 1740 by the brilliant 20-year-old French entomologist Charles Bonnet, who found that without male intervention, the body of a single female aphid may be found to contain the bodies of her progeny, and, not only that, these progeny contained the even smaller germs of *their* progeny: two generations of individuals telescoped together in the body of a single creature.

The products of parthenogenesis are always identical clones, and always female. The advantage of parthenogenesis is that it allows creatures that live in difficult environments to capitalize on any temporary or ephemeral resource that might come along, by breeding rapidly for as long as their good fortune lasts — much more rapidly than could be achieved by sexual reproduction. For aphids, that resource is your roses, or your tomatoes. Many other creatures apart from aphids are known to have the facility to become parthenogens, if they have need for it. Several species of salamander and one kind of lizard have populations that are parthenogenetic, but there are no known naturally occurring cases of parthenogenetic birds or mammals.

In general, parthenogenesis is a temporary solution. Large populations of genetically identical individuals are much more vulnerable to disease than populations of sexually reproducing ones, the reason being that sex is the best known way to generate variety quickly. Variation is so important in the never-ending war between organisms and disease that it is possible that sex first evolved as a way in which organisms could keep one step ahead of disease-causing bacteria. For this reason, populations are rarely parthenogenetic for more than a few generations. Sooner or later, a parthenogenetic population will become extinct, or males will reappear.

As always with the living world, there are exceptions. Ponds and puddles of all sizes are often inhabited by exquisitely beautiful microscopic creatures called rotifers. Because ponds are among the most ephemeral habitats — a few warm days and they may vanish entirely — rotifers tend to be parthenogenetic more than is the case with animals in general. This is taken to the extreme in the Bdelloidea, a whole group of parthenogenetic rotifers that seems to have been in existence for tens of millions of years, and in which no males have ever been seen.[5] Apart from being a suitable icon for feminism, the existence of at least 360 species of bdelloid rotifer proves that an all-girl lifestyle need not be a barrier to evolutionary diversification.

There is nothing about parthenogenesis that precludes Orcs from this mode of reproduction. The small, fluctuating populations of Orcs living in marginal habitats such as mountains where food is scarce (after all, parties of edible Dwarves and Hobbits do not turn up every day) would favor this mode of reproduction. The case of the bdelloid rotifers shows that different and distinct species can exist even when all the organisms concerned are entirely parthenogenetic: each colony of Orcs in the Misty Mountains is its own species, living in its own isolated and ephemeral pond. The only problem with this scenario is a small jolt of perspective, nothing more. Despite the unvarying, male-centered terminology associated with Orcs; for all that Bolg is the 'son' of

Azog; for all the relentless militarism; and for all the 'boys' and 'lads' of orkish banter, if Orcs were parthenogenetic, then *every Orc would be female*. The Great Goblin would be a Queen, and Azog and Bolg would in fact be mother and daughter.

The strangeness of this perspective shift may be softened by the knowledge that this kind of militaristic (if not militant) feminism exists in the real world. Packs of hyenas may be ruled by a single dominant female whose dominance is achieved, in part, by aggression caused by excessive concentrations of male hormones. These hormones effect a transformation on her body as well as her behavior; the genitalia of dominant female hyenas become masculinized, and the birth canal shrinks to such an extent that her ability to reproduce as a female is compromised.[6] Although hyenas are not parthenogenetic, reproductive females do look startlingly male down to the details of their genital anatomy. By the same token, Orcs could *still* have been female, despite the evidence of one's own eyes were one unfortunate enough to have caught one without its trousers.

Parthenogenesis could be the solution to the enigma in which Orcs appear either to be sexless or, at least, to have no females. This scenario is consistent with everything we know about Orcs, and also points up Tolkien's depiction of males and females in every species of note *except* Orcs.

Dwarves are particularly notable in this context. Reporting that Dwarves are a peculiarly strange folk (*Rings* F), Tolkien observes that Dwarves are commonly thought to have no women, a case of a male-only species born directly from rock. The Dwarves, though, report that female Dwarves do exist, but are much rarer than males, and in any case look similar to males because they, like male Dwarves, are bearded. This curious note (see 'Concerning the Dwarves' in *The Later Quenta Silmarillion, HOME* XI) shows that Tolkien was prepared to adopt a very broad view of physical form when discussing non-human species. If neither male nor female Hobbits were bearded, then it is entirely plausible that both sexes of Dwarves had beards – and that all Orcs were biological females even if they appeared to be male.

7. ARMIES OF DARKNESS

In the previous chapter I discussed the biology of Orcs in their wild state, governed loosely, if at all, by Sauron or Morgoth. The next task is to examine the greater problem of their origin and the nature of their relationship with the dark powers, a matter that preoccupied Tolkien from the first drafts of *The Book of Lost Tales* in 1917 right up to notes made in the last years of his life.

The question is simply stated, if rather less easy to solve: were Orcs manufactured by Morgoth; pre-existing creatures that had been perverted into their orkish state; or a combination of both? The problem is one of 'sub-creation,' Tolkien's central and abiding preoccupation. Tolkien's view, informed both by his Catholicism and his knowledge of the processes whereby myths and legends are made, was that sub-creation — the creation of languages, artifacts, imaginary worlds of fable, and so on — is a fundamentally human trait, which as humans we have a duty to pursue. Scientists, as sub-creators, have an obligation to explore the universe for the same reason that has driven scholars since time immemorial, that is, to glorify the Name of the Creator.

But all privileges carry responsibilities, and sub-creation is no exception. Sub-creations do not belong to their makers, because both the maker and the made are ultimately held on leasehold from the Creator (*Letters* 153). For this reason, sub-creators have a duty to their creations to hold them lightly at their fingertips, and not guard them too jealously as if they were theirs, and theirs exclusively: that would be the prerogative of the Creator alone. When Ilúvatar — 'The One' — created Elves and Men, he left them to order their own affairs, not even allowing the Valar very much power to govern them. The summoning by the Valar of the primeval Elves from Middle-earth to Valinor was an act that the Valar themselves came to regret, as an act of unwarranted meddling (see *The Istari* in *Unfinished Tales*).

The jealous hoarding of the Silmarils by Fëanor was a more open case of subcreative disobedience, echoing Melkor's earlier rebellion in his wish to take over the whole world as if it were his dominion to order as he would, denying it to others. Melkor, however, wanted to create life itself, but would soon have found that this was impossible. A Vala — that is, a demiurge — could create many things, even the bodies of living beings, but these would behave as automata unless granted the spirit of life by Ilúvatar. This is made clear in the legend of the creation of the dwarves by Aulë, the Vala with special responsibility for the fabric of the Earth, who wished to share his delight in its wonders with others of like mind (*The Silmarillion*, chapter 2). In what must rank as a version of the *Pinocchio* story, Aulë created the Dwarves, but he found that they were incapable of independent thought or movement, behaving as if they were Aulë's own remote limbs — in effect, robots. When Ilúvatar chided Aulë for his hubris, the miscreant Vala was so contrite that Ilúvatar allowed the dwarves to have independent life, provided that they 'awoke' in Middle-earth after Ilúvatar's own creation, the Elves.

The Aulë story makes it very clear that had Melkor sought to create Orcs in the same way that Aulë had made dwarves, then his sub-creations would have behaved as drones, incapable of independent thought or movement, as mindless as any other machine. Nevertheless, I should like to propose that Melkor, being a Vala far more powerful than Aulë, might have been able to create counterfeits of life even more convincing than Dwarves without any help at all from Ilúvatar, for all that they were automata. Could these counterfeits have behaved, for all practical purposes, like sentient, incarnate, 'real' beings? How could we have even told the difference?

The manufacture of Orcs in some underground cavern looks like the best and simplest way of creating an Army of Darkness quickly, with the least fuss. This appears to be the implication of a statement (*Rings* A) that the race of Uruks made its first appearance outside Mordor around 2745 of the Third Age – but this could simply mean that these creatures had existed in Mordor long before, but had not been seen abroad until that time. However, some early notes suggest that Tolkien was thinking very much along the lines of outright manufacture for his armies. In the earliest version of *The Fall of Gondolin*, the city is besieged not by the Orcs, trolls, and other organic creatures we are used to from, say, the Pelennor Fields or Helm's Deep, but by vast, articulated, fire-breathing, machines (*HOME* II). It may not be a coincidence that Tolkien wrote *The Fall of Gondolin* in 1917, the same year as the Battle of Cambrai, the first engagement to have involved tanks, something that must have struck Tolkien as quite appalling. In later versions of *The Fall of Gondolin*, and in everywhere else, the combatants were all living creatures, but still, here and there, a sense of artifice peeps through. The Orcs are dehumanized and mechanized, and throughout Tolkien's writing is an ambiguity about whether Orcs, balrogs, and so on were manufactured by Morgoth, just like those primeval fire-breathing engines of war, or were sentient beings that had fallen into Morgoth's snares and had become corrupted.

As I have already noted, the behavior of Orcs in *The Lord of the Rings* shows that they *can* think for themselves, always to their own detriment and in a way that confounds and frustrates the plans of their masters, whether Saruman or Sauron. This is not to say that mere manufactured automata are unable to exercise their own judgment, and in so doing might be hard to distinguish from sentient beings. This perennial theme was engendered by Mary Shelley in *Frankenstein* and taken up very successfully by, among many others, Philip K. Dick in *Do Androids Dream of Electric Sheep?*, filmed by Ridley Scott as *Blade Runner* — and in a series of pulp robot stories by Isaac Asimov.[1] At issue here is whether the judgment of local Orc commanders is motivated by self-interest to the extent that it runs counter to the will of their superiors. Of course, to expect that Sauron's war machine is as efficient as *Blitzkrieg* is to miss the point: Tolkien's own experience of active service on the Somme would have shown him that wars are fought just as much by the muddle, attrition, stupidity and waste of the many, as by the heroic acts of the few. As I have noted with reference to the Rohirrim in contrast to the Orcs, the contrast between these two modes of warfare (with the heroism of the few explicitly associated with virtue) is a recurring theme in *The Lord of the Rings*. However, to expect that Orcs were, in origin, manufactured automata — even if, once let loose in the mountains, they resorted to biological means of reproduction — might be to expect too much of the evidence: especially as Tolkien posits and discusses another point of view to the creation of Orcs as organic robots.

That alternative, endlessly fretted and picked over in the course of the decades-long development of *The Silmarillion*, was that Orcs were, in origin, captured Elves. The endless indecision is amply illustrated by notes from the *Annals of Aman* (in *HOME* X) in which Tolkien writes that Orcs are enslaved Elves, broken by Morgoth and bred. However, notes added to this suggest that Orcs should not, in the end, have such an Elvish derivation. Deepening the confusion is a constant alternation between the idea of creation and corruption. The word 'made' is emended to 'bred,' 'spawn' is changed to 'children' and so on: the later *Quenta Silmarillion* (*HOME* X) has Orcs 'made' in mockery of Elves, in a passage referring to the manufacture of monsters more generally.

In the published version of *The Silmarillion*, at least, some of the first Elves, walking in the starlit forests of Middle-earth, were spirited away by the dark powers, ruined, corrupted, and made into the first Orcs. This solution seems neat but falls apart on even the slightest examination. As Tom Shippey notes in *The Road to Middle-earth*, Morgoth would have had to have gathered an awful lot of Elves to have created his armies, so much that someone would surely have worked out what was going on and done something about it. But had these brutalized Elves, now Orcs, been forced to breed, they would have created not Orcs but more Elves, which would have had to have been brutalized in their turn. This savage cruelty might have been well within the powers of Morgoth, but creates obvious problems for the generation of Orcs once outside the immediate purview of Morgoth, Sauron, or Saruman. One way of getting round this would be to have engineered the reproduction of Orcs so that they would always breed true, as Orcs.

One might also speculate that natural selection might play a part, by selecting the more Orkish of any offspring of Elves brought up in the harsh environment of an Orc colony. However, this might require longer periods of time than available in Tolkien's legendarium, and would certainly imply vast wastage of Elvish bloodstock along the way: the waste inherent in natural selection is usually taken for granted.[2]

However, these were not the problems that troubled Tolkien about this scenario. Tolkien was less worried about the biology of turning Elves into Orcs than about the attendant theological conundrums. Elves, like all incarnate beings, had souls, which were the gifts of Ilúvatar to the body, which acted somewhat as a 'house.' Even though the body might be altered, from Elf to Orc, the spirit would have remained inviolate, beyond the reach of the tormentor. Whenever an Elf died, its spirit went to the Halls of Mandos in Valinor: had Orcs been simply Elves remade, the death in battle of multitudes of Orcs would soon have filled the halls with many disembodied and very angry spirits. Nowhere in the discussion of the Halls of Mandos in the published *Silmarillion*, or even on the detailed discussion of Elvish mortality in late, unpublished works such as *Laws and Customs among the Eldar* and *Athrabeth Finrod Ah Andreth* (both in *HOME X*) are Orcs mentioned in this context. Being a Christian and a Catholic, Tolkien was also taken to task about whether Orcs, having been corrupted through no fault of their own, were not incapable of salvation.[3] If they were, the wholesale slaughter of unsaved Orcs in the cause of good is made even more tragic. (On a smaller scale, we see quite clearly that Gollum is not completely beyond the reach of salvation, and very nearly repents, making his death all the more affecting.)

In a miscellany of remarkable, late writings collected as *Myths Transformed* (*HOME* X) Tolkien sought to recast many of the fundamental aspects of his invented world. Turning once more to the perplexities of Orkish origin, Tolkien came up with two radical, completely satisfactory and overlapping solutions. Orcs were not derived from Elves at all, but from corrupted *men* (something that Treebeard suspects of Saruman's *Uruks* in *Rings* III,4), engineered *animals*, or a combination of both. This scheme (or class of schemes) cuts decisively through the Gordian knot. If Orcs are, in origin, Men, the problem of what to do with Elvish souls goes away, given that the fate of human souls is unknown in the compass of the invented world — nobody, not even the Elves, knows the destiny of human souls, so the overcrowding of Mandos is no longer a problem. By the same token, this solution resolves the conundrum of the creation of automata that defy their masters, because Orcs can, indeed, be derived from pre-existing creatures rather than manufactured *de novo*. Asserting that animals are somehow involved finesses the problem, as corrupted animals, even if raised to a near-human level, are still animals — soulless, behaving in many ways like instinct-driven automata, and creatures for which the agonizing problems of damnation and redemption are irrelevant.

The origin of Orcs as men also allows for the various gradations between Man and Orc described in *The Lord of the Rings*, from the various human villains with an Orkish streak — Bill Ferny, the 'Squint-Eyed Southerner' at Bree, the footpads and ruffians of *The Scouring of the Shire*; to the explicitly Man-Orc hybrids that are Saruman's Uruks. These varying degrees of admixture, modulated by history, genetics, and evolution, would account in part for the pronounced racial variation between Orcs. Finally, the specific association of Orcs with Men allows Tolkien to comment, through his stories, on the Fall of Man, the human condition, and the dehumanizing effects of power, progress, technology, and warfare on ordinary people.

In terms of science, these various grades of Orc-human mixture can be read as a savage critique on evolution itself — or, at least, the view of evolution as 'progressive,' leading to inexorable improvement in form and function. This is the view of evolution that would have been current in the first half of the 20th century, and most especially between 1900 and the end of the Second World War, encompassing Tolkien's most productive years as a writer. I have shown elsewhere that this view of life is profoundly antithetical to what we now understand of the Darwinian model of evolution by natural selection, and has indeed been exposed as illogical by theorists working from the 1950s onwards.[4]

Natural selection, the basis of Darwinian evolution, is simply an easy way to refer to what happens when the forces of nature 'select' the most suitable variants of any population, according to the prevailing circumstances of the environment. Because variation is inherited, the result will be that the overall complexion of the species will change, tracking the environment. Over time, hunters will evolve sharper eyes to track the hunted, which will evolve better ways to flee or to hide.

However, there are two clear messages here about what evolution is *not*. The first is that creatures do not acquire helpful features as they grow older, such that they can pass them on to future generations. Natural selection simply allows the 'fittest' to survive, and the others to die, in the same way that a cat breeder in search of cats with furrier coats will breed only from the furriest cats in any litter. Although furry cats tend to have furrier kittens, this is not a consequence of furriness having been acquired in a single generation: whereas it is true that the children of a self-made millionaire tend to be born rich, this mode of inheritance is not applicable in biology.

The second important message is that natural selection has neither memory nor foresight, nor any sense of purpose or will: it works only in the infinitesimal here and now, and so can tell us nothing about the grand sweep of evolution over millions of years, still less *why* certain features evolved. Therefore, common statements such as 'humans stood upright *so that they could* free their hands for holding tools,' 'fishes evolved legs *so that they could* walk on land' or 'birds evolved feathers *so that they could* fly' are profoundly unscientific, at least in the context of natural selection. However, the popularity of the view of evolution as progressive, with a will and a purpose and a bent towards improvement, endures in the public mind: witness the regular use in advertising of the canonical parade from apes, through ape-men, to upright humans, culminating in the latest consumer product with a tagline such as 'Move Up to the Next Step in Evolution.'

The mistaken view of evolution as inevitably progressive and improving owes much to the 19th-century German movement known as *Naturphilosophie*, in which all organic forms were seen as manifestations of a cosmical 'striving' towards Man as the most perfect form. As the prominent *Naturphilosoph* Lorenz Oken (1779-1851) put it:

> What is the animal kingdom other than an anatomized man, the macrocosm of the microcosm?[5]

Although *Naturphilosophie* contains many good things, the effects of its application to evolutionary biology by the biologist and writer Ernst Haeckel (1834-1919) have been disastrous, in that they have confused generations of people about the nature of Darwinian evolution. Haeckel was Darwin's number one fan in Germany and did much to spread the word of *Darwinismus* there in the late 19th century. But because of his upbringing in a scientific *milieu* deeply dyed by *Naturphilosophie*, Haeckel welded old ideas of progression onto Darwin's theory of evolution by natural selection — itself profoundly antithetical to progression — and created a monster that has regrettably endured in the public imagination. Even today, science news stories bearing headlines about 'missing links' owe their underlying philosophy more to Haeckel's misperceptions than to contemporary science. However, back in the 1930s, Tolkien would have been familiar with the anthropological manifestation of the Haeckelian view of science, then scientific orthodoxy, in which the human species was graded into races of 'lower' or 'higher' status: the view that led to Nazi theories of race.

Through the subcreative lens of Middle-earth and applying this widely held (but mistaken) view of evolution as progressive and improving — with humanity occupying a station between Apes and Angels — we can imagine a parade of forms starting with Moria Orcs and proceeding in sequence through Mordor Orcs, Uruks, Squint-Eyed Southerners, Dunlendings, Bree-Men, Rohirrim, Men of Gondor and finally High Numenoreans aspiring to transcendent Elfhood. It is important to remember that Tolkien would never have implied this to mean a scheme of evolutionary transition. For one thing, Elves and Men are separate creations in Tolkien's world, and Orcs are corruptions: so, if anything, the arrow would have pointed the other way, towards degradation rather than elevation.

However, adherents of *Naturphilosophie*, and the other pre-Darwinian biologists who came up with the *scala naturae* — the 'ladder of nature' — with *Urschleim* at the bottom and humans at the top, did not intend for their plan to be interpreted in any evolutionary way, either. That interpretation was not made before Haeckel. The *scala naturae* was simply a static plan of the Divine organization of nature, with every creature occupying its ordained station, and Tolkien's world should be interpreted very much in that light. Nevertheless, some pre-Darwinian thinkers thought it acceptable for creatures to strive for betterment in true nature-philosophic style, just as they can also slide down the greasy pole into barbarism. In that sense, if evolution occurs at all in Tolkien's world (and the biology of Orcs suggests that it can), its mechanism owes something to the ideas of French savant Jean-Baptiste de Lamarck (1744-1829).

Lamarck's views are very similar to those of the German nature-philosophers: in his treatise *Philosophie zoologique* (1809), Lamarck proposed that animals could move up the ladder of nature by the exercise of a primal, creative force, or *besoin* ('need'). Creatures that achieved some goal *in their own lifetimes* could pass these acquired traits on to their offspring: here was a view of evolution *precisely* equivalent to the inheritance of riches by the offspring of self-made millionaires. The canonical example of Lamarckism in action is of short-necked giraffes actively lengthening their necks while straining to reach the topmost leaves of trees, and passing on their lengthened necks to subsequent generations. Lamarckism is generally discredited today, but it is worth noting that in the period when Tolkien was working out his mythology, Lamarckism was much more popular than it is now. It was, in fact, seen as a viable alternative to Darwinism, which went through a bad patch around 1900 as a result of theoretical and practical difficulties that remained unresolved until the integration of evolutionary theory with the emerging science of genetics in the late 1930s.[6]

Prominent intellectuals such as Samuel Butler (1835-1902), author of the anti-Darwinian satire *Erewhon* (1871) and *The Way of All Flesh* (published posthumously in 1903) promoted Lamarckian thought, and as late as 1916, Lamarckism was the basis for a popular student textbook on zoology in England.[7] A Lamarckian (rather than a Darwinian) model of evolution in Middle-earth solves one of the problems raised by the production of Orcs from corrupted Elves, in that an Elf would pass its degradation *directly* onto any offspring without further intervention from Morgoth, and would subsequently breed true as regards its acquired, Orkish traits.

The Lamarckian model works just as well in the scenario that Orcs are really ruined Men, rather than Elves, but the result is more deeply moral, and puts some of the motivations of the characters into perspective. Consider: in this scheme, Men, Orcs, and half-

Orcs can descend into a bestial state much more easily than Aragorn can conquer his own inner weakness and strive to regain the glory of his longer-lived forefathers — a problem compounded by the long, slow decline of Numenorean powers, as evidenced by the decreasing life spans of the Dunedain. With this in mind, Aragorn would have looked at the Orcs and realized that for him, if not for the Elves, there existed the potential for him to fall an awfully long way, something that is made clear to him by Elrond in *The Tale of Aragorn and Arwen* (*Rings* A): a parallel to Frodo's mixture of horror and pity when he looked at Gollum and realized that this is how he, too, could end up.[8] Aragorn's victory, therefore, is made far more important because he is a Man, not an Elf, and furthermore because he is a Man of the highest possible lineage, and therefore with the most to lose of any creature in Middle-earth. No lesser man runs the same risks as Aragorn; no higher Man exists. Elves, in this scheme in which Orcs and Men are connected by a greasy pole, exist in a state of grace, beyond all risk.

The slow degradation of thousands of years is one thing, but how could human beings ever be persuaded to *interbreed* with Orcs — created by whatever means — as Tolkien explicitly states that they can? Tolkien has two possible solutions to this distasteful dilemma, both consistent with a progressive, possibly Lamarckian view of evolution. First, as explained in *Myths Transformed*, he has Morgoth progressively reduce Men, over generations, to a bestial level, so that they would not be too particular about their partners: second, Tolkien has Morgoth create Orcs by elevating mindless creatures of unknown origin almost to the level of sentience. These strategies compound what Tolkien might have viewed as the sinful miscegenation of adjacent but separate rungs on the *scala naturae*. They also have clear parallels in literature.

Perhaps the most obvious antecedent is *The Island of Dr. Moreau* (1896), perhaps H.G. Wells' finest novel, in which the hero, Prendick, finds himself on a desert island inhabited by the notorious vivisectionist Moreau, his drunken assistant Montgomery, and a population of what Prendick initially takes to be congenitally ugly indigenes. It dawns on Prendick that this ugliness is not natural, but a product of sadistic experiments on human beings. Prendick is, however, mistaken, as Moreau later explains: the natives are not vivisected humans, but animals, raised by the vivisector's art to the near-human state. This is simpler to achieve than one might think, says Moreau, because the line between the rational human and the non-sentient animal is much thinner than is usually assumed:

> Very much indeed of what we call moral education is such an artificial modification and perversion of instinct: pugnacity is trained into courageous self-sacrifice, and suppressed sexuality into religious emotion. And the great difference between man and monkey is in the larynx ... in the incapacity to frame delicately different sound-symbols by which thought could be sustained.

If the parallels between Moreau and Morgoth are obvious, those between Moreau and Saruman are even closer: Moreau, like Saruman, has a mind wholly devoted to mechanism at the expense of spirit, is in the business of breeding a semi-sentient bestial race in the manner of Uruks — and he even has the hapless Montgomery to play the part of Gríma Wormtongue. There is, however, an even closer parallel, and that is between Moreau and Sauron. Like the beast-people, the trolls (*Rings* F) were animals, bred by Sauron to the verge of sentient thought, such that they were able to speak albeit in a somewhat debased fashion.

One of the most intriguing points to emerge from *The Island of Dr. Moreau* is Prendick's initial assumption that despite their deformities, the natives are human, simply because they can speak, and use that speech to worship Moreau as a God and express what seem to be deeply human emotions about their own place in the Universe. The most poignant scene in the book is where Prendick witnesses a ritual of the 'Beast People:'

> The dark hut, these grotesque dim figures, just flecked here and there by a glimmer of light, and all of them swaying in unison and chanting: 'Not to go on all-Fours: *that* is the Law. Are we not Men?

This religion is, however, a sham, no more than animal instincts articulated as speech. After Moreau's death, the Beast People gradually lose the power of speech and revert to the status of animals. The point that Wells (an atheist) is making is that once blessed with language, even a soulless automaton can create a passable imitation of such quintessentially human activities as religious expression. If that is the case, then it might be very hard to distinguish, in any practical way, between an automaton and a sentient individual.

Another reference to the transformation between man and beast, and the near-illusory nature of the barrier that divides them, can be found in a much earlier work, *Gulliver's Travels* (1726), Jonathan Swift's satire on the human condition. You would think that there could be no greater contrast between the works of Wells, a utopian atheist with a seemingly boundless optimism about the progress of the human species; and those of Swift, a deeply dystopian cleric whose novel is a hymn to misanthropy. Yet there are many parallels between *Gulliver's Travels* and *The Island of Dr. Moreau*, and some interesting contrasts.

In his fourth voyage, Gulliver finds himself in the land of the Houyhnhnms, a race of intelligent horses. The Houyhnhnms are courteous hosts, but are inclined to be rather cold and rational. Their otherwise perfect society is continually disrupted by animals called Yahoos, whose behavior is in complete contrast to the civility of the Houyhnhnms. Where the Houyhnhnms are decorous, peaceful, and modest, the Yahoos are disgusting, violent, licentious, and with no more language than unpleasant grunts and gestures. Gulliver is horrified to discover that were it not for his clothes, language, and a veneer of civilization, he, too, would be classed as a Yahoo. This is made clear when his Houyhnhnm hosts ask Gulliver to describe the society of his own country, as honestly as possible, for the Houyhnhnms are incapable of deceit and are inclined to take things literally. It is here that Swift delivers his scathing satire on the squalor and corruption of English society and politics: in his examination of his own condition from the dispassionate perspective of the Houyhnhnms, Gulliver comes to realize that even a civilized Englishman such as himself can be distinguished from a Yahoo only by fine words and a powdered wig. The Houyhnhnms banish Gulliver from their country, a verdict with which Gulliver can only concur.

In their examination of the human condition, *Gulliver's Travels* and *The Island of Dr. Moreau* both conclude that human beings, for all their pretensions of civilization, are separated from the animal world not by some great divide but by a line so thin that it can be crossed almost without our being aware of it. This can best be seen by the dysfunctional reactions of the narrators once returned to normal human society, in England.[9] *Moreau's* Prendick reports that:

> I could not persuade myself that the men and women I met were not also another, still passably human, Beast People, animals half-wrought into the outward image of human souls; and that they would presently begin to revert, to show first this bestial mark and then that.

Swift, speaking through the returned Gulliver, goes beyond reportage to excoriating moral condemnation:

> My reconcilement to the Yahoo-kind in general might not be so difficult, if they would be content with those vices and follies only, which nature hath entitled them to ... but when I behold a lump of deformity and diseases, both in body and mind, smitten with *pride*, it immediately breaks all the measures of my patience; neither shall I be ever able to comprehend how such an animal and such a vice could tally together.

The difference between the two works is one of perspective. In *The Island of Dr. Moreau*, Wells confronts us with animals that are initially taken to be human, to the extent that they practice religious ritual and try very hard to be human beings. Wells is making the point that what we think of as specifically human activities, such as speech and even religion, are nothing special, but merely elaborations of motivations that already exist in the animal world.

In *Gulliver's Travels*, we see the complete opposite: the Yahoos are biologically indistinguishable from humans, but rightly condemned as animals because of their appalling behavior, a consequence of their flawed natures. Salvation, if it is indeed possible, must lie in our own acceptance of our bestial natures, an acceptance which requires a certain modesty. The Beast People are raised to be simulacra of humans by science: the Yahoos have degenerated of their own accord.

In both cases, speech is central. Prendick is fooled by the speech of the Beast People into thinking that they are akin to humans; Gulliver is fooled by the Yahoos' *lack* of speech into believing precisely the opposite. It should be no surprise that Tolkien regarded the presence or absence of speech as the mark of sentience. The name that the Elves gave themselves — *Quendi* — means 'the speakers'(*Rings* F).

Tolkien's Orcs present an interesting amalgam of Yahoos and Beast People. Orcs are like Yahoos in that they are human beings reduced to degenerate savagery. It is true that Orcs are capable of language, but it is of a primitive and barbaric kind. On the other hand, the direct involvement of the dark powers in the creation of Orcs (something on which all the various hypotheses of Orc origin can agree) adds a *Moreau*-like dimension. Orcs are Beast People in the sense that they have been actively made or modified from other creatures, and, like Beast People, the Orcs display the appearance of sentience.

Can this question of sentience be used to resolve the origins of Orcs as either automata or sentient beings? Can it be used to demarcate Orcs from the other 'speaking peoples'? If the examples of *Gulliver's Travels* and *The Island of Dr. Moreau* are any guide, the answer to both questions is 'no.'

Because human beings are the only extant creatures to use language, we tend to judge the humanity of unknown entities on whether they are capable of speech, even though the real world repeatedly shows that this distinction is not nearly so clear as we should like. We are often fooled into thinking that such non-human, non-sentient things as parrots and voicemails have human attributes *simply* because they use language (a confusion wherein lies a rich seam of comedy.) Research into artificial intelligence has shown that it is possible for humans to conduct apparently meaningful conversations with computers without knowing that their interlocutor is a machine: computer pioneer Alan Turing showed long ago that the distinction between man and machine might be very hard to draw.

Even creatures we have good reason to suspect are sentient — ourselves — can spend appreciable amounts of time acting as if they were not. As Jack Cohen and Ian Stewart show in their book *Figments of Reality*, the human 'mind' is not switched on all the time, but 'coalesces' only when it is needed. Many activities, such as sleep, driving a car along a well-known route, or playing a familiar tune on the piano, can be undertaken without much in the way of conscious thought. Indeed, one could argue that we would be profoundly impaired were we forced to concentrate on every passing second of our lives, a condition explored by Jorge Luis Borges in his short story *Funes the Memorious*, about a young man disabled by total recall.

In *Myths Transformed*, Tolkien suggests, with reference to Orcs, that 'talking' is not necessarily a sign that a spirit inhabits the body, that is, of sentience. Despite their chatter, therefore, Orcs were beasts raised to man-shape, trained like dogs and horses, and taught to speak parrot-fashion.[10]

This has an important consequence for our understanding of the place of Orcs in Tolkien's world: it shows that Orcs might be able to act independently and in contravention of their orders, and still be soulless, non-sentient automata. But when the chips are down, Orcs revert to the status of mindless ants. Tolkien's description of the behavior of Sauron's troops when their master is defeated is very telling – he describes them as running mindlessly hither and thither, as colonial creatures might do were their queen to be removed (*Rings* VI,4). In this climactic moment of *The Lord of the Rings*, the Orcs – whatever their biology, whatever their origins – revert to mere machines of war, the toys of Morgoth as described so long before in *The Fall of Gondolin* (*HOME* II), written while Tolkien was experiencing at first hand the birth and immediate consequences of mechanized warfare.

8. THE LAST MARCH OF THE ENTS

The Ents are among the most remarkable and original of all Tolkien's creations. The word 'ent' comes from an Old English word simply meaning 'giant,' but Tolkien's transformation of these gigantic creatures into the 'shepherds of the trees' was a master-stroke.

Walking trees do not occur in real life, with the possible exception of banyan trees. These tropical trees propagate by sending aerial roots down to the ground from constantly spreading branches. These roots grow into secondary trunks, so that after a while a single tree can form its own ready-made forest. As new trunks form, old ones may die off, so that the tree as a whole shifts around. This, however, is rather different from the more vital tree-giants that Tolkien had in mind.

Literature, like science, has relatively little to say about intelligent and possibly mobile vegetation. Tolkien would, of course, been familiar with the universal folk-wisdom of trees; the Norse conception of the world as a tree, Yggdrasil, and that the Norse Adam and Eve were created as trees, *Ask* and *Embla* ('ash' and 'elm.') Folklore tends to paint trees as indifferent to human concerns, or even malevolent, a trend taken to extremes in more recent times in the form, for example, of the terrifying protagonists of John Wyndham's *The Day of the Triffids* (1951): the triffids being giant, mobile, poisonous plants that take over the world, as things so often do in Wyndham's fiction.

The triffids came too late to have influenced Tolkien's own conception of Ents, but it is possible that at least some of the roots of the Ents might be found in Olaf Stapledon's novel *Star Maker*, published in 1937. We know that Tolkien knew Stapledon from the casual reference in *The Notion Club Papers* to the 1932 novel *Last Men in London* (see Chapter 1). But in *Star Maker* we meet creatures that seem obvious precursors for Ents, as Stapledon's star-voyager explores sun-drenched worlds inhabited by the 'Plant Men,' described as 'gigantic and mobile herbs.' However:

> To say that they looked like herbs is perhaps misleading, for they looked equally like animals. They had a definite number of limbs and a definite form of body; but all the skin was green, or streaked with green, and they bore here or there, according to their species, great masses of foliage ... In general those that were mobile were less generously equipped with leaves than those that were more or less sedentary.

This short passage, together with others making it clear that the Plant Men were sentient beings, gives a fair prescription for an Ent, as well as hinting at that gradation of character described by Treebeard from fully mobile Ent to the more sedentary individuals that had become 'tree-ish.' *Star Maker* precedes by a whisker the first mention of the 'giant Treebeard' in the early drafts for *The Lord of the Rings*, so it is possible that Stapledon's Plant Men might have set seed in Tolkien's mind for them to have contributed to what eventually came to be Ents. It may be noteworthy that Treebeard started as a regular giant: his more overtly arborescent character emerged as Tolkien wrote, some time in 1939.[1]

Whatever the literary precursors for Tolkien's Ents, we see in these huge, thoughtful, extremely ancient and unarguably alien creatures Tolkien's preoccupation on the one hand with trees (see, for example, *Letters* 165) as well as with his greater, all-pervading theme of loss. Although Tolkien's love of trees and forests have found favor with ecological activism, the condition of the Ents reflects recent scientific thinking on the circumstances of species confronting inevitable extinction.

By the time Merry and Pippin run into Treebeard in *The Two Towers*, the Ents are coming to the end of a long, slow decline that will, in Treebeard's own view, culminate in their extinction, sooner or later. To Treebeard, the reason is clear: all the remaining

Ents are male, and so can no longer reproduce. The female Ents (the 'Entwives') have long since left the scene, and nobody knows where they might have gone. They had crossed the Anduin to colonize the Brown Lands, which by the end of the Third Age have been desolate for an indeterminate period (*Rings* II,9; III,4).

In addition, Ents are very long lived. Treebeard is hailed as the oldest living thing in Middle-earth (*Rings* III,5), with the arguable exception of Tom Bombadil. Like all long-lived creatures, Ents would reproduce only very slowly, even were the Entwives still among them. These facts, taken together, illustrate the inevitability of the Ents' extinction. It has been a long time since a young Ent (an 'Enting') has been seen in Fangorn, the forest home of the Ents, and the remaining Ents are becoming ever more tree-ish. The dwindling cadre of Ents resembles a collection of World War I veterans who, meeting annually on Remembrance Sunday, record their inevitable depletion and one-way slide into decrepitude and death. It is possible that as a combat survivor himself, Tolkien meant the Ents to convey precisely this impression. However, there are other, more subtle clues about the biology of Ents that illustrate their predicament — clues that have been validated by recent work on ecology, particularly of ancient species that stand as relics of vanished ages.

We are offered very few explicit facts about the details of Entish reproduction, but we can assume that it was a kind of hybrid between the kinds of reproduction seen in animals and in plants. Like higher animals in general, Ents are either invariably male or female, and not both at once. Many plants are hermaphrodite, with individuals having both male and female organs. Most garden flowers are hermaphrodite, bearing clusters of pollen-bearing anthers (the male organs) and a single, central pistil (the female organ.) Sometimes, a species of plant will have hermaphrodite plants as well as male-only plants, but this is unusual.

Many plants can also reproduce vegetatively: that is, they can clone themselves by sending out suckers. Gardeners can exploit this tendency by taking cuttings. However, the tendency among long-lived plants, especially trees, is that the sexes are separate, and vegetative reproduction is limited or absent.

If Ents are considered as animals, or as 'speaking people,' their reproduction is unremarkable. After all, with the possible exception of Orcs, all the people we meet in Middle-earth, no matter how strange they may be, appear to be either male or female and reproduce in a manner akin to that of human beings. However, if we think of Ents as *plants*, their style of reproduction becomes significant, because Ents reproduce not as *any* plants, but *specifically* as long lived, woody plants, like the trees they tend. Moreover, it is impossible to take a cutting from an Ent and root it in a convenient tub of potting compost. If this were so, then the departure of the Entwives would not have been a problem, and the tragedy of the Ents' inevitable extinction (very much in accord with Tolkien's preoccupation with loss) would be entirely absent.

There are yet other pointers to the desperate straits of Entish reproduction. Many species whose numbers have undergone a drastic decline suffer from reproductive problems because they are, in terms of genetics, far too 'samey.' For example, the few cheetahs that remain in the wild are genetically very similar to one another. This is thought to have contributed to their further decline, although this is controversial.[2]

Genetic sameness means that creatures are inclined to become rather inbred, a condition that tends to bring inherited disorders out into the open. This phenomenon, known as 'inbreeding depression,' is a common problem in zoo animals, and is familiar from human societies whose traditions favor a choice of mates from within the same society, and which, therefore, tend to suffer from the disproportionately large incidence of diseases that are otherwise rare in the population as a whole. Examples are not hard to find. The Amish of Pennsylvania tend to suffer from bipolar disorder (manic depression) far more often than is the case with people generally; European Jews suffer from a developmental disorder called Tay-Sachs disease; and people in some Mediterranean countries suffer more from a blood disorder called thalassemia than is the case with people at large. As Jared Diamond reports in *The Rise and Fall of the Third Chimpanzee*, each one of the many isolated tribes of New Guinea suffers from its own, idiosyncratic tendency towards one disorder or another, be it albinism, leprosy, or something else.

The Ents, however, seem to be suffering the opposite problem. Their curse is not limited variation, but far too much of it. As Merry and Pippin note from the Entmoot which they attend, each one of the Ents has its own physical idiosyncrasies. On first meeting a new and alien species, one might expect to have a hard time telling the difference between individuals. This is taken to the extreme by the Elves of Rivendell, who can hardly tell the difference between Aragorn (a Man) and Bilbo (a Hobbit) on the grounds that, to them, one mortal looks much like another, as if they were two sheep in a flock. In contrast, Merry and Pippin are immediately struck by the individual variation of the Ents — in height, color, skin texture, and even the numbers of fingers and toes. Of course, Tolkien is trying to show how each of the Ents is akin to one particular species of tree or another, such as a rowan or a beech, but this variation has a consequence that has direct relevance to the imminent extinction of the Ents. There are real-world cases in which ancient species of long-lived individuals, somewhat out of place in the modern world, resort to extremes of individual variation in a last and desperate bid to survive.

One particularly telling case is that of *Dedeckera eurekensis*, a rare perennial shrub only found in clumps in the Mojave Desert of California.[3] Believed to be a relic from earlier, more equable times, individual *Dedeckera* plants can live to more than 140 years, and cling grimly on in a harsh, arid climate to which they have not evolved. Like English colonials in the tropics who go out in the mid-day sun, *Dedeckera* persists in maladaptive strategies such as flowering in midsummer when other desert plants hunker down. Despite extensive, vigorous flowering, very few *Dedeckera* fruits set seed, and virtually all of these abort. Like the Ents, *Dedeckera* cannot propagate by cuttings or suckers, so the plants must reproduce through seeds, however desperately, to survive.

Why should these shrubs reproduce so poorly? Why do they persist in their adaptation to a habitat that has long since given way to desert? It turns out that, possibly like Ents, *Dedeckera* plants differ enormously in their genetics, from one plant to the next. One would expect that this would promote hybrid vigor, but there comes a point — and this is accentuated in long-lived species that are very rare — when the accumulation in each parent of harmful genetic mutations becomes too great for the offspring to bear, and it simply aborts. It is possible that this enormous degree of variation is the last, desperate throw of a plant that is trying to evolve to suit its new conditions, until its own innate capacity for variation is exhausted, and at the cost of drastically reducing its ability to reproduce. In an interesting detail, *Dedeckera* is a notable part of the flora of the Last Chance Mountains: like the Ents, *Dedeckera* is on its last march towards its final doom.

This syndrome of desperate variation seems to have afflicted the Ents. Even were the Ents once again able to meet and cross with Entwives, this degree of variation means that very few viable Entings would be born (or germinated), and that the population of Ents would probably not be able to grow fast enough to avoid annihilation as a result of a second, killer factor. That factor is habitat loss.

Large-scale deforestation is a feature of Middle-earth as it is in the real world. Tolkien explicitly states the Old Forest and Fangorn are just two, small remnant patches of a forest that extended in one, unbroken swathe from the Shire to Dunland. It is implied that disease and warfare were important factors in the general degradation of the land, though changes in climate might also have played a part. The hand of Man, however, is explicit: in *Unfinished Tales* (see *The Port of Lond Daer*, Appendix D of *The History of Galadriel and Celeborn*) Tolkien states that the forests of Enedwaith and Minhiriath were cut down in the Second Age largely as a result of the demands of the Numenorean ship-building industry.

When we meet him at the end of the Third Age, Treebeard talks wistfully of woodland journeys taken in the distant past, including (in the context of a song) to the forests of Nimbrethil and Ossiriand in the land of Beleriand, a region almost entirely destroyed at the end of the First Age, more than 6,000 years earlier. Merry and Pippin ask the obvious question of why Treebeard no longer makes journeys like this, in view of the pressing need to find the Entwives: but for Treebeard, the time for wandering passed long ago, and the remaining Ents are content to stay in Fangorn to await their fate rather than look for pastures new. The stay-at-home nature of Ents, of course, can only hasten their extinction, so why does Treebeard view the prospect with such equanimity?

My guess is that, old and wise as he is, Treebeard knows that the fate of the Ents is inevitable, even were the Ents far more numerous than they evidently are. Ecologists have come to similar conclusions for real-world forests: it is now strongly suspected that the clearance of forests affects the fate of the creatures within depending on whether they were 'dominant competitors' or not.[4] The surprise is that dominant competitors tend to be more prone to extinction than seemingly less successful species, and that this extinction — this doom — becomes inevitable after a certain amount of habitat clearance, even if the position of the species seems secure at the time of census. In other words, forest clearance marks down many successful species for inevitable extinction, even if that event is many years in the future.

Some explanation is needed for terms such as 'dominant' and 'competitor.' Although the Earth is inhabited by many millions of species, ecology sees only two — competitors and dispersers. Competitors succeed by standing their ground in certain places where they may be common, or even dominant; dispersers succeed by spreading themselves thinly over a wide area, without necessarily becoming dominant in any one place. At first sight, dominant competitors in a patch of untouched forest look to have a fairly secure future. That is, if such a species occupies ten per cent of a forest, and a random ten per cent of that forest is clear-cut, the presence of sites in the remaining ninety per cent should ensure survival for a competitor that is sufficiently abundant. However, it turns out that the destruction of a random ten per cent has the same eventual effect as selectively destroying *precisely* that ten per cent occupied by the species: that is, an inevitable extinction, even if deferred for decades or centuries.

Applying this scenario to the destruction of the great forest of which Fangorn is merely a remnant, we can see that Ents are dominant competitors. They dominate the ecology of the patch of forest in which they live, but are disinclined to venture beyond it. Given that Fangorn survives as but a small, relictual remnant of a much more extensive habitat, it may well be the case that the inevitable doom of the Ents was wrought long before, when the axes of the Numenoreans brought down the woods of Eriador, splitting the continuous forest into separate northern and southern patches.

There is, however, a tiny chink of hope for the Ents. Early in *The Lord Of The Rings*, the Hobbits discuss possible sightings of walking trees in the Northfarthing (*Rings* I,2). Could these have been the Entwives? And how had this sighting have slipped the minds of the Hobbits, when confronted with the Ents? Tolkien never follows this up, and this isolated event remains an intriguing loose end.

And what of the flock tended by the shepherds of the trees? The trees themselves are treacherous, even dangerous, to the extent that tree-herds are necessary protection for any two-footed beings that innocently stray under the eaves of the ancient forests. In Fangorn, it is the Ents who have taken on this necessary policing role. The Old Forest adjacent to the Shire is in many ways very similar, and it is here that Tom Bombadil rescues the Hobbits from the depredations of Old Man Willow, who rules the Forest from his station on the bank of the River Withywindle. Bombadil makes it clear to the Hobbits that this single, ancient, and rotten-hearted tree is largely responsible for much of the evil reputation of the Old Forest as a whole. Many of the trees have come under his influence, through invisible threads woven through the air and laid under the ground (*Rings* I,7). It is notable that in the initial description of the Old Forest, there are many descriptions of trees but almost none of individual species, more or less until willows are mentioned on the banks of the Withywindle. It is as if all the trees have become subdued by a single power – Old Man Willow — their individual identities or specific affiliations completely subordinated.

This kind of tree-to-tree influence is far less mystical than it seems, for it has a strong real-world parallel. It may be found among the roots of all trees, whether in great forests or just at the bottom of the garden. Most people spend their lives entirely ignorant of it, even though life on the Earth's surface would be very different in its absence — perhaps even impossible. It is a kind of fungus, collectively the mycorrhizae. The mycorrhizae bind together the trees on Earth in a wood-wide web that governs the history of a forest and the lives of all the trees within it.

Like most fungi, mycorrhizae live virtually all their lives underground. When we think of fungi we tend to imagine mushrooms and toadstools, but these are simply the small, visible fruiting bodies of much larger underground creatures. Even though fungi constitute the most massive organisms known,[5] their bodies are extremely diffuse and extensive, being made of microscopically thin tubes called hyphae. In general, fungi get their living from the decomposition of corpses, whether of animals or plants. Mycorrhizae, on the other hand, live in close association with living things, forming close associations with the roots of living trees and other green plants.

These links are symbioses — that is, both the fungus and plant gain from the link to the extent that each cannot survive without the other. These plant-fungal symbioses are hardly rare curiosities. Indeed, it is thought that most, if not all plants have mycorrhizal partners. Some mycorrhizae are specific to particular plant species, whereas other kinds of mycorrhizae can support many different plant species. It has been suggested that plant life on land could only become securely established once plants and mycorrhizae had established this symbiotic way of life, around 400 million years ago.[6]

The tree provides food for the fungus, in the form of sugars distilled by the tree's own leaves from carbon dioxide, water, and sunlight, in the process of photosynthesis. The fungus, in return, uses its hyphae to extend the root system of the tree deeply and widely into the soil, scavenging important minerals and possibly fending off micro-organisms that might damage the tree.

It is in the soil that the networks of adjacent mycorrhizae meet, linking different trees to one another. Although trees seem like isolated steeples to us, this is merely an illusion derived from our above-the-ground parochialism. Much competition and trading happens between our feet, in a wood-wide web of mycorrhizal threads that has the potential to connect all the trees of a forest into a single unit. It is now known that trees use mycorrhizae to exert their influence on each other. For example, recent work has shown how hemlocks (*Pseudotsuga menziesii*) and birch (*Betula papyrifera*) exchange nutrients through the mycorrhizal network, but that the balance is tipped slightly in favor of the hemlocks, especially if they are in danger of becoming shaded by the birches.[7] The malignity of Old Man Willow, spreading throughout the Old Forest, from root to root, from tree to tree, was very far from being as fanciful as Tolkien might have imagined.

9. O FOR THE WINGS OF A BALROG

Balrogs are the demons of the ancient world, the fearful flame-wreathed servants of Melkor Morgoth, the first Dark Lord. However, what keeps Tolkien fans awake at night is neither the fiery swords, nor the whips; the shadow nor the flame, but whether Balrogs had wings or not. Here I'd like to add my two cents to the argument. I think that the answer, on balance, is 'not': at least, not any wings that were capable of flight as we conventionally understand it.

The principal reason is that Tolkien is never explicit about the presence of wings in Balrogs, even though he goes into great detail about wings in other flying and fantastical creatures — and that one should, as a matter of course, take Tolkien at his word, where possible. The second reason is based on aerodynamics. That is, if the wings were to have been used in flight (which is what the scanty hints that survive tend to lead one to believe) then they would have to have been ridiculously huge — *so* huge, that not to have mentioned them at all would have been strange indeed.

The first time that most Tolkien readers will meet a Balrog is on the bridge of Khazad-Dûm in *The Fellowship of the Ring*, where the two references to wings (in *Rings* II,5) are contradictory. This Balrog — the Balrog of Moria — is initially described as a 'shadow' in the midst of which is a humanoid form, perhaps greater than the height of a Man. It has a fiery mane, and holds a flaming sword and a whip. When the Balrog gains the Bridge, the shadow about it reaches out like a pair of enormous wings. These wings are plainly described as the appurtenances of shadow, and they could be just that — shadows cast when a great dark shape is illuminated from below, for example by the light cast by flames burning in a fiery chasm. The second reference seems more explicit, stating that its wings were spread from 'wall to wall.' However, given the first reference, these could still be wings of shadow, a cloak of smoke neither more substantial nor capable of flight than Count Dracula's cape, and in any case not meant to be taken literally.

The case against wings in Balrogs is very strong. Balrogs go right back to *The Book of Lost Tales*, but none of the many references to them before *The Lord of the Rings* even hints that they had wings or were actually capable of flight. This is in complete contrast to dragons, where Tolkien makes a clear distinction between wingless dragons (such as Glaurung in the *Silmarillion*) and flying, winged dragons (such as Ancalagon, again in the *Silmarillion*, or Smaug in *The Hobbit*.)

The vagueness about Balrog wings also contrasts with the very clear description of the wings of the flying steeds of the Nazgûl. In *The Return of the King* these creatures are described in some detail as featherless, with wings made of webs of skin stretched between horned fingers, in other words, somewhat like giant bats (*Rings* V,6). However, in their great size, somewhat reptilian mien and, rather more notably, their description of having survived from some primeval age, these creatures are rather more like pterodactyls (more properly, pterosaurs) the order of flying reptile that became extinct at the same time as the dinosaurs, 65 million years ago.[1]

As an interesting but possibly tangential aside, one of the protagonists in *The Notion Club Papers* (in *HOME* IX) is described as having written a book of poems entitled *Experiments in Pterodactylics*: given that *The Notion Club Papers* and *The Return of the King* were written just one before the other, I think that pterosaurs, rather than bats (still less Balrogs) was what Tolkien might have had in mind for the flying steeds of the Nazgûl.

However, this idea is modulated, if not contradicted, by Tolkien himself. Responding to a direct question from a reader about whether these creatures were meant to be pterodactyls, Tolkien wrote that this was not, in fact, his explicit intention (*Letters* 211). Nevertheless, the creatures were — again in his words — 'pterodactylic' — in that the allusion offered a way of portraying them as the last relics of a vanished prehistory. The allusion to pterodactyls, therefore, was meant to conjure up general images of antiquity and extinction, rather than to mean that the mounts of the Nazgûl were meant to be taken literally as pterosaurs.

Tolkien also notes rather playfully that his creation is probably less mythological than the reconstructions offered by paleontology: indeed, Tolkien's reading of the state of paleontological thought in the 1950s is sharply accurate.[2] To emphasize the ambiguity, Tolkien refers to the creatures himself as 'Nazgûl-birds' (*Letters* 100). This ambiguity should not be seen as authorial indecision or vacillation, but a deliberate strategy that maintains the mystery of the story, and therefore its excitement and allure. To my mind, it would have been a great pity had Tolkien simply admitted that 'Nazgûl-birds' were meant to be pterodactyls. Where would the fun be in that? Of course, one could say the same thing about the wings of Balrogs: part of the enjoyment lies in *not* being quite sure. On a larger canvas, the fun of scientific enquiry lies in the realization that discoveries never achieve definitive conclusions but only open up further vistas for our exploration, rather than in the definite, precise details demanded by obessional fans, or by people who think, mistakenly, that the business of science is to provide concrete answers.

Back to the Balrogs. Although Tolkien refers to Balrogs repeatedly as belonging to his miscellany of monsters, they feature strongly in only two episodes in the *Silmarillion* tradition. The first is the episode in which Balrogs rescue Melkor from Ungoliant. The second concerns the prominent role of Balrogs in the Fall of Gondolin.

Melkor, returning to Middle-earth after the Rape of the Silmarils, is almost overcome by the spider Ungoliant, but is rescued by Balrogs lying in wait for their master's return. In the *Silmarillion* as it stood in 1951 (*HOME X*, the later *Quenta Silmarillion*) the Balrogs are reported as passing 'with winged speed' to their master — a provocative phrase, but which need only refer to the speed of the Balrogs rather than their physical means of achieving it. In any case, the passage just cited becomes still more enigmatic, saying that the Balrogs rushed to Melkor as a fiery tempest (and similar wording appears in the *Silmarillion* as published), with no mention of wings at all, even as adjectives. In a thematically similar passage in Appendix A of *The Lord of the Rings*, a Balrog is described as 'flying' from Thangorodrim, but without an explicit description of wings, this might be no more than flight in the sense of rapid and desperate escape, a use as metaphoric as 'winged speed.' In any case, Balrogs were Maiar, beings of the same supernatural order as wizards and Sauron, and so could have had abilities to transcend those of the prosaic and corporeal. One presumes that they could have sped through the air much like Superman, who, it has been claimed, flies faster than a speeding bullet, with nothing more aerodynamic or substantial than a flimsy red cape.

Balrogs lead the final assault on Gondolin, but in all the intense detail of the battle for that city (*HOME* II), in which the hero Ecthelion dies in the act of slaughtering Gothmog, Lord of Balrogs, and where Balrogs are described as being heavily armored and twice the height of an Elf (and remembering that the Elves at that time were conceived as people of heroic stature), no mention at all is made of Balrogs having wings or being able to fly. When the refugees from Gondolin (again in *HOME* II), led by the brave Glorfindel, are ambushed by a Balrog in the mountain passes, the monster is described as 'leaping' onto the sorry convoy, not once, but twice — and wings are never mentioned. This is particularly significant in that the eagles of the Crissaegrim join the battle at that point, and rescue Glorfindel from the chasm into which he and the Balrog had fallen, locked in battle. Balrogs remain resolutely wingless in early drafts for *The Lord of the Rings* itself (*HOME* VII) where the Balrog of Moria is described as no taller than a Man but similar in every other way to the creation as published, except that wings are not mentioned at all, not even as metaphoric or ephemeral shrouds of shadow.

The shape and attributes of Balrogs are established almost from the first: demons of fire, humanoid in shape, and ranging in height between that of a Man and twice that, with the characteristic fiery sword and whip. But nowhere at all are they *explicitly* described as having wings, in contrast to the clear descriptions of wings in the other flying creatures described by Tolkien. Because of this, the wings of the Balrog of Moria, if they have any substance at all, are wings of smoke and shadow rather than flesh and bone.

This begs the question of why the Balrog of Moria in Peter

O FOR THE WINGS OF A BALROG

Jackson's film adaptation of *The Lord of the Rings* had wings. These are seen quite clearly in the opening sequence in *The Two Towers*, in which Gandalf and the Balrog plunge into the chasm of Khazad-Dûm. One could argue, however, that these wings are ornamental, especially as the creature makes no attempt to fly out of trouble, either in the chasm or later, on the peaks of the Misty Mountains.

Not that this purely literary argument is likely to convince diehards in the Balrogs-have-wings camp, so I shall have to resort to experiments in aerodynamics — and, to an extent, pterodactylics. A traditional way of proving a contention in mathematics or logic is to admit the opposite, and follow it to its remorselessly illogical conclusion. In which case, let us suppose, for a moment, that Balrogs really *did* have wings. What would they have been good for? Could they have been used for flight? My answer to that one is, again, probably not, or, at least, not for the purposes of rapid airborne pursuit, which is all that is allowed from the tiny hints that Tolkien gives us.

Compared with birds, Balrogs were big. The largest birds to fly today are the Kori bustard (*Ardeotis kori*) of Africa, with a mass of 16 kg; the wandering albatross (*Diomedea exulans*) has a mass of 10-12 kg, and the South American condor (*Vultur gryphus*) is 14 kg. These were matched in the very recent geological past by Haast's Eagle (*Harpagornis moorei*) of New Zealand (up to 13 kg), a raptor that preyed on the giant moas (*Dinornis*), a group of flightless birds akin to the ostrich, and which, like their flying predator, are all now extinct.

The Ice Age of the Americas saw the evolution of some very large, vulture-like birds of prey akin to condors, known as teratorns. Merriam's teratorn (*Teratornis merriami*) had a mass of around 13 kg and a wingspan of more than three meters, comparable with that of a modern condor. It preyed on animals trapped in treacherous asphalt seeps just off what is now Wilshire Boulevard in downtown Los Angeles. But the biggest flying bird ever discovered lived about six million years ago in Argentina. This bird, *Argentavis magnficens*, was the mightiest of the teratorns, with a mass of 80 kg and a wingspan of six meters.[3] This creature was just the right size to have carried a few Hobbits in its talons, or even a wizened wizard on its back, and makes a perfect model for the heroic eagles such as Gwaihir or Thorondor who always seem to turn up just in time to save the day. Roosting in the pinnacles of what is now the Andes, the eyries of *Argentavis* would have been homes fitting for the Eagles of the Lords of the West.

Although many kinds of enormous, non-flying birds are known to have lived — one, the superficially ostrich-like *Phorusrhacos*, stood 2.5 meters tall and had a truly fearsome beak in a skull the size of that of a horse — it is not known whether any birds existed larger than *Argentavis* that could have flown.

The problem of flight is not so much one of staying aloft — it is getting up there to begin with. The crucial parameter is something called 'wing loading,' which describes the mass of the creature as a function of its wing area. A large creature can take off, provided it has a big enough area of wing to support its weight, and can reach a certain minimum take-off speed. This take-off speed depends on the wing loading, and it turns out that *Argentavis* would have had to have been moving at around 40 kilometers (25 miles) per hour before it would have generated enough lift to get airborne. This is why large birds, such as swans and geese, flap vigorously and have a long run-up to get themselves into the air. However, it is doubtful whether a bird with a six-meter wingspan could have flapped its wings while standing on the ground, simply because its legs would be too short to allow the wings much flapping room: albatrosses, which have very long wings and relatively short legs, are very poor at taking off from the ground for this reason. In any case, it is worth remembering that the only birds to achieve 40 km/h on the flat are specialized, flightless, long-legged runners such as ostriches and rheas: the legs of teratorns are stout and stocky and not built for fast sprinting.

But airspeed is relative: all that matters is an airflow of sufficient speed is moving across the wings, irrespective of whether the bird is running pell-mell towards it or just standing still in a brisk airflow. All a large bird needs to do to take off is to run against a stiff breeze, or, if the wind is strong enough, simply face the wind and open its wings, and it can take off from a standing start. Condors take off in precisely this way. It is possible that, in life, *Argentavis* roosted in the Andes and hunted on the Pampas to the east. Stronger easterly breezes blew across the region then than is the case now, because the Andes, being young, still-rising mountains, were lower than they are today. These breezes would have allowed the birds to have taken off from the Pampas with little difficulty, once they had landed to scavenge or hunt. The breezes that blow across the Pampas have weakened since, perhaps explaining why condors still exist, but the larger teratorns do not.

Take-off is less of a problem for large birds living in trees, cliffs, or mountain peaks. All they need to do is fall off their perches, open their wings, and gravity will do the rest. It is not for nothing that large birds such as eagles, condors, and albatrosses live in mountains or on high, rocky islands. This, incidentally, is another reason why, if the Balrog of Moria had functional wings, it would easily have been able to use them to fly out of trouble in the chasm of Khazad-Dûm incident, or, later, by launching itself from Durin's Tower.

As birds grow larger, their wings must expand in area at a greater rate than the mass they are required to lift. This is why you see small birds (but not large ones) flying successfully with small, stubby wings. But large wings impose an additional penalty — they require greater physical effort to flap up and down. Large birds flap more slowly than smaller ones, and this imposes certain restrictions on how they live. Condors and albatrosses do not flit from peak to peak entirely by their own muscular exertion. Instead, they gain lift from rising currents of air called thermals; from air moving up mountain slopes; and, in the case of albatrosses, from air deflected upwards from the surfaces of waves. Thermal soaring, in which large birds such as vultures cruise from thermal to thermal, sometimes for hundreds of kilometers without a single flap, is probably how *Argentavis* got around.

Another consideration that affects large birds is wing shape, or what engineers call 'aspect ratio,' that is, the ratio of wingspan to the mean chord, or breadth of the wing. Wings with a low aspect ratio are deep and relatively stubby, whereas wings with a high aspect ratio are long and narrow. When air moves across a wing, it tends to break up and become turbulent at the trailing edge, creating drag. This effect is minimized in narrow, high-aspect-ratio wings, because the air can pass all the way across the wing before becoming turbulent. Fliers with narrow, high-aspect-ratio wings, such as albatrosses, or sailplanes, can fly very efficiently, exploiting the slightest updraft of air with minimum loss of height, and can also circle very tightly at low speed. Vultures, on the other hand, have broad, low-aspect-ratio wings, but get around the problem of drag by spreading their trailing-edge flight feathers, creating trailing-edge slots. A bird with a low aspect ratio will sacrifice lift for speed: this is an advantage for a soarer winging its way between thermals, and for a bird of prey that wishes to lose height very quickly, such as when it stoops on its prey from a great height.

We do not know if larger birds than *Argentavis* ever flew, because the maximum size of birds is determined by considerations of lifestyle, environment, and metabolism as much as by aerodynamics. However, it is doubtful that any birds grew as large as the largest pterosaurs, such as *Quetzalcoatlus*, which lived in Texas in the Late Cretaceous Period around 80 million years ago.

The smallest pterosaurs, distant reptilian relatives of both birds and dinosaurs, were no bigger than sparrows. The largest, however, had wingspans of the order of ten or eleven meters — as much as a small plane, and twice as great as *Argentavis*. However, the skeletons of pterosaurs were constructed with extraordinary lightness. Birds are light for their size because many of their bones are hollow, and some are fused into rigid, airframe-like structures. Pterosaurs took this tendency to the extreme, and the result was enormous wings, very low wing loading, and, possibly, high aspect ratio — although this last is controversial, because paleontologists disagree about the wing shape of the largest pterosaurs.

Large pterosaurs would probably have been expert soarers, taking advantage of the tiniest air currents, maneuvering on the weakest, narrowest thermals by circling within their own wingspans, and being able to take off by opening their huge, hang-glider-like wings into the mildest breezes.[4] They would have flapped only when cruising close to the ground or the sea, when air reflected off the surface would have given them an extra boost. In this respect, the depiction of the huge Nazgûl-birds in the Peter Jackson film adaptations of *The Lord of the Rings*, flapping very slowly and in general only when they are close to the ground, is very accurate indeed, for that's what any large flying animal would do in the same set of circumstances.

So why are there no large pterosaurs today? Or if not pterosaurs, why have no animals evolved to fill the niche of giant, living sailplanes? The answer could simply be environmental. In the

Cretaceous period, the polar regions were not glaciated as they are today. Currents of warm, equatorial seawater ran right round the globe, so that contrasts of temperature were far less than they are nowadays. This would have meant that winds were, in general, light: perfect for a sailplane, but not for anything more robust.

Things changed when Antarctica moved across the South Pole, preventing seawater access to keep it warm, and when the northern continents moved to isolate the Arctic Ocean from warmer seas. The result was an increase in the contrast in temperature between poles and tropics, which created much windier weather than the large pterosaurs, adapted for flying very slowly in mild conditions, could have tolerated. Accustomed as they were to nothing stronger than a light breeze, the ensuing tempests would have blown the pterosaurs out of the skies. Continental drift, then, not the arrows of keen-sighted Elves, would have done for the pterosaurs in the end, even if they had not perished at the end of the Cretaceous Period 65 million years ago, quite suddenly, along with the dinosaurs.

Where does all this leave the Balrogs? Let's look at Balrog aerodynamics from the very few clues Tolkien left us. Tolkien said that a Balrog was of approximately humanoid shape, and between one and two times the height of a man. Let us compromise and say it was about one-and-a-half-times as high, or about three meters tall. Without any other clues about the Balrog's build, we have to take Tolkien at his word and assume that if he likens it to a human in shape, it might have been like a human in build, and had no unusual bird- or pterosaur-like weight-saving features such as hollowed bones. Using a standard ideal height-weight chart as a rough guide, a humanoid this tall would have a mass of around 140 kg: that is, getting on for twice the estimated mass of *Argentavis*.

However, as described in *The Bridge of Khazad-Dûm*, the Balrog had a wingspan that stretched from 'wall to wall.' This is as close as Tolkien gets to a description of Balrog wingspan, even were this feature not considered to be merely a metaphor for shadow. However, we have no idea of the dimensions of the hall concerned, except that it was large. We might hazard a guess that the hall was, say, 30 meters across, but this estimate is, frankly, as good as any other. In any case, the span was likely to have been very much larger than the Balrog's actual height, effectively dwarfing it. So, unless they looked like Dumbo's ears, these wings would have had to have been quite narrow.

Assuming, therefore, an extremely high, albatross-like aspect ratio of about 20, the wing area of a Balrog would have been 30 x 1.5 = 45 square meters, and its wing loading would have been 140/45 = 3.1 kg/m². This would make for a very light flier, with a proportionate wing loading far less than the very heavy Canada goose (loading 20 kg/m²) or even the relatively light peregrine falcon (9.1 kg/m²). *Argentavis* had an estimated wing loading of 11.4 kg/m², which is lower than one might expect for a soarer of its size, and even lower than expected for a marine soarer such as an albatross. Even compared with this, the wing loading for a Balrog seems incredibly small. Decreasing the aspect ratio by broadening the wing, while keeping wingspan the same, would have made the wings even more extensive, decreasing the wing loading even further.

From this it is simple to calculate the minimum airspeed that a Balrog would have needed to get airborne. This speed is given by dividing the wing loading (measured not in kg/m² but in pascals, which is a measure of weight rather than mass) by a constant determined by wing shape, and taking the square root of this quantity. As a rule of thumb, the wing-shape constant is about 0.9kg/m³ for a well-designed wing. If we assume that the wings of Balrogs were so designed, the minimum take-off speed works out as 5.84 meters per second, the equivalent of 21 km/h or about 13 miles per hour: this rates a moderate breeze on the Beaufort scale, enough to raise dust and move small branches. This is half the estimated take-off speed of *Argentavis* — but remember that we have adopted a relatively narrow, high-aspect wing for a Balrog. If we were to reduce the aspect ratio (that is, broaden the wing) while retaining the same 'wall-to-wall' span of 30m, we would increase the wing area still further and reduce minimum take-off speed. Adopting an eagle-like aspect-ratio of about 9, Balrog wing area becomes 99 m², loading becomes a tiny 1.4 kg/m³, and minimum take-off speed is just 3.9 meters per second, which translates as 14 kilometers or just 8 miles per hour, a mild breeze strong enough to be felt on the face, and to rustle a few leaves.

In light of these calculations, the wing loading of a Balrog was such that it would have been able to take off by extending its wings over the updraft caused by an energetic fire, which would have been just the thing for Thangorodrim, if not for within the confines of Moria. However, aerodynamics would have constrained it to have behaved like a glider: it could never have flapped these enormous, sail-like wings, which would have been precisely the wrong kind of wing for the rapid interception implied when Tolkien alludes to Balrogs in flight, most notably, the tempestuous, winged pursuit to rescue Morgoth. Say what you like about sailplanes, but I do not think that one can ever describe their motion as 'tempestuous,' for any tempest, or even any rapid motion, would have crumpled a flying Balrog like an umbrella in a hurricane.

But all these arguments are as nothing when you examine the great size of a Balrog's wings, at least given the meager hints offered by Tolkien. Wings that big would have been a Balrog's most prominent feature, so it is strange that Tolkien never mentioned them until the later drafts of *The Lord of the Rings*, and even then in such ambiguous terms. Or perhaps the wings were *so* huge that any self-respecting Balrog would have been — well — too *embarrassed* to have drawn attention to them? Yes, I think I'll have seen just about everything, when I see a Balrog fly.

There is a possible let-out clause, however: that the wings of Balrogs were not really wings, but shadowy extensions of the body designed to make them seem even larger and more frightening. In other words, just like the cape of Count Dracula. And despite Tolkien's protestation at the existence of Disney (*Letters* 13), I see a strong resemblance between Balrogs and the enormous, very scary winged creature depicted in the 'Night on the Bare Mountain' sequence in *Fantasia* (1940) and the 'Firebird' sequence in *Fantasia 2000* – pointing up the contrast between fire and shadow that is the essence of Balrog, as well as displaying wings which, while terrifyingly huge, were too big to have been aerodynamically feasible, at least for the purposes stated by Tolkien.

10. SIX WHEELS ON MY DRAGON

Tolkien was fascinated by dragons from an early age, but he soon ran into a problem — that is, northern legend contains rather few dragons from which to draw inspiration (*Letters* 122). There's Fafnir, fought by Sigurd the Volsung, and there's the dragon in *Beowulf*, and that's just about it. However, there is just enough to sketch the essential features of these remarkable creatures. They are very large, serpentine reptiles, some of which are capable of flight.[1] So far they resemble Nazgûl-birds, but Tolkien's dragons have three other distinguishing features — they breathe fire; they can cast spells on their victims when within close range; and the winged ones have six complete limbs, that is, four legs and a pair of wings. Some of these features have the potential to be problematic, but scientific research, some of it fairly recent, can shed light on the possibilities and potentialities of life as a dragon.

Fire-raising is less of a problem than one might imagine. All that is required is that the dragon can synthesize a liquid that can be stored in a gland, but which ignites easily when squirted out into the air. The range of possible liquids is great, and reptiles (and many other animals) are known to engender all kinds of unpleasant substances when excited or threatened. Snakes, for example, produce venoms, glandular secretions injected through fangs, and some beetles emit hot, noxious liquids when irritated. The bombardier beetle (*Stenaptinus insignis*), for example, synthesizes a mixture of hydrogen peroxide and hydroquinone which, when the beetle is threatened, it releases into a 'combustion chamber' where enzymes provoke the two substances to react with each other, producing the toxic substance benzoquinone and a great deal of heat. The boiling hot liquid is then squirted with fair accuracy at the unfortunate assailant.

Given the complexity of this system, one might easily imagine the biological synthesis of a compound that ignites spontaneously when ejected forcefully into the air. One candidate might be the organic solvent, diethyl ether. This is a colorless but sweet-smelling liquid that produces a copious, highly flammable vapor. Because it vaporizes so easily, diethyl ether is notable for its flashpoint, that is, the lowest temperature at which it produces enough vapor to be ignitable when in contact with air. Diethyl ether's flashpoint is a chilly -45°C, much lower than most common laboratory chemicals.

Once in contact with air, diethyl ether does not need to get into a flame to start burning fiercely: all it needs is a warm stovepipe, or even the static electricity generated by moving the liquid from one vessel into another. A dragon could keep reservoirs of liquid ether tightly bound in airtight sacs, perhaps, like venom glands in snakes, evolved from salivary glands. Forcible, muscular movement of the liquid through the blood-hot mouth cavity and across the teeth would be sufficient for ignition and the production of an impressive yellow-orange flame. Because diethyl ether produces so much vapor, a little of it can go a long way. An ether fire burns anything organic, and the fact that it is less dense than water — and does not mix with it — means that it cannot easily be extinguished. If the fires of Smaug burned with the aid of ether, the thatch and wooden pilings of Lake Town in *The Hobbit* would never have stood a chance.

Ether has the organic structural formula $CH_3CH_2OCH_2CH_3$, and is easily made in the laboratory by warming ethanol (ordinary grain alcohol) in the presence of sulfuric acid. The acid 'dries' the ethanol, fusing two ethanol molecules together to create a molecule of diethyl ether, removing the elements of water.

$$2CH_3CH_2OH \rightarrow CH_3CH_2OCH_2CH_3 + H_2O$$

Dragons would do this by harnessing the power of various microbes, which would form ecosystems in the glands of dragons in the same way that microbes live in profusion in our guts, or the guts of cows, producing, as by-products, the gas methane (which, as all nocturnal pranksters know, is satisfyingly flammable.) Many micro-organisms produce ethanol as a by-product of digesting sugars (this is the basis of wine-making). Some bacteria, on the other hand, oxidize sulfur, or sulfur-containing minerals such as iron sulfide ('iron pyrites' or 'Fool's Gold'), to produce sulfuric acid. Ether might be a natural by-product of such a system, and it is possible to imagine microbes found solely in the ether-glands of dragons, which have enzymes designed to maximize the production of ether from ethanol at the expense of other byproducts.

Keeping a vapor-producing, fat-dissolving substance such as ether in a gland could, however, pose mechanical problems. The gland would have to blow off excess vapor occasionally (perhaps accounting for the constant low-level fuming of dragons, even when not spewing flames). In addition, some way would have to be found of preventing (or, at least, minimizing) the escape of ether through the lining of the gland, intoxicating the dragon itself. One would have to assume that the dragon could tolerate a concentration of ether in its own tissues that would kill anything else. Nevertheless, it is possible to imagine ether glands lined with various sulfate minerals, which tend to be insoluble and settle readily from solution, although this would put the dragon at risk of discomfort or disease were such minerals to block the ducts between the glands and the mouth.

Why ether, rather than — say — kerosene, or even benzoquinone, as in bombardier beetles? In addition to its properties as a flammable solvent, ether is also a potent anesthetic. Just a whiff of ether is enough to make one feel drowsy, and a little more can knock you out. Ether (as well as hypnotism) might explain why people, when in the presence of dragons, are either stunned into a helpless torpor, or lose their minds for shorter or longer periods. In *The Hobbit*, Bilbo is conscious of the need to be particularly wary when in the presence of dragons, and the story of Turin in *The Silmarillion* (and its many earlier versions in the *History of Middle-earth*) turns on the ability of the dragon Glaurung to cast spells on the hapless human cast. Glaurung freezes Turin into motionlessness on the bridge at Nargothrond: later, the dragon causes Turin's sister Nienor to lose all memory of her life up to that point.

A particular problem with winged dragons is that they have too many limbs for a regular reptile – that is, wings in addition to four fully-developed legs. All other land vertebrates never have more than four limbs, arranged in two pairs, fore and hind, and the wings of flying vertebrates are invariably modifications of the forelimbs. Is it possible to go against this pattern, to create a genuinely six-legged vertebrate, in which one of the three pairs of legs is converted into wings, leaving two pairs for walking?

Before going any further, it is possible to subvert the problem entirely, by supposing that the wings of dragons are not modified wings at all, but derive from extensions of the body wall, supported by modified ribs, as found in the modern gliding lizard *Draco*. However, the maneuverability of dragons in the air, and their degree of control, suggests that their wings were more than passive gliding membranes, but capable of a power stroke, and this implies the kind of musculature associated with fully developed limbs.

If we are to confront this problem directly, two solutions present themselves. The simplest is to suggest that an extra pair of wings is the product of some kind of mutation, bred to fixation in the underground laboratories of Morgoth. 'Sports,' in which creatures are born with supernumerary limbs, are well known, especially in frogs exposed to potent disruptors of embryonic development such as vitamin A.[2] These sports are the results of disruption to the activities of a remarkable group of genes responsible for setting out the body-plan of organisms, ensuring that every organ in the body would develop in its proper place with respect to every other organ.

The first inkling that such genes existed came in the work of the pioneering English geneticist William Bateson (1861-1926). In a book called *Materials for the Study of Variation* (1894), Bateson sought to explore the underlying character of natural variation on which evolution depended, but which was not, at that time, understood. *Materials* is largely a bestiary — a catalog of 'sports' or monsters, examples of creatures with more or less than the number of organs that we should normally expect of the endowment of nature; or of creatures with normal organs, but misplaced. Thus in *Materials* we see an endless parade of creatures with abnormal numbers of claws, legs, wings, teeth, bones and so on; of moths with too many wings, and bees with little legs growing out of their heads in place of antennae. Bateson realized that there was a special category of monster in which organs such as legs or wings developed perfectly normally, but were abnormally duplicated, or in the wrong place, and he coined the word 'homeosis' to refer to this phenomenon.

More than eighty years passed before the genetic basis of homeosis was understood. Working at the California Institute of Technology in Pasadena, Edward Lewis found that homeotic transformations were linked to a distinctive cluster of genes — at least in the fruit fly *Drosophila melanogaster*, the experimental animal of choice in genetic laboratories for more than a century.[3] Disruptions to one or other of this gene cluster leads to distinctive mutations in which legs develop in the place of antennae, or flies with an extra pair of wings.

This last mutation is particularly interesting from the point of view of the anatomy of dragons, and to understand it we must learn a little about the anatomy of flies. The middle region of the body of any insect, behind the head and in front of the abdomen, is called the thorax. The thorax consists of three well-defined segments, each of which bears a pair of legs. In the fly, the second segment bears a single pair of wings. Insects such as dragonflies, which are more primitive (in evolutionary terms) than flies, have two pairs of wings. In flies, however, the second pair of wings — sprouting from the third thoracic segment — has evolved into a pair of tiny drumstick-shaped structures called halteres, which whirl around like gyroscopes and this explains why flies have such perfect balance and maneuverability. The mutant flies studied by Lewis have a pair of wings growing from the second segment, as usual; but a second pair of wings emerging from the third segment, in the place normally occupied by halteres. This mutation is more than just the addition of wings. The *entire* third thoracic segment, rather than developing as normal, has developed as a homeotic *duplicate* of the second segment.

This simple example suggests that homeotic genes may have had an enormous part to play in the evolution of animal body form. After Lewis's pioneering work with flies, an explosion of biological research has revealed that most animals more complicated than a shapeless sponge have homeotic genes arranged in distinctive 'clusters' that govern the body form of the animal concerned. This is as true in vertebrates, such as ourselves (and dragons) as it is in fruit flies. However, apart from the presence and function of homeotic genes, flies and vertebrates otherwise develop in very different ways. Although the role of homeotic genes in the development of vertebrate limbs is becoming progressively better understood, geneticists have not been able to create – say – a six-legged laboratory mouse, in which the shoulder and upper-chest region has been duplicated. Vertebrates and flies are sundered by more than a billion years of separate evolutionary progress, so their modes of development cannot be compared that closely.

Ultimately, the reason why it might not be easy to create a six-legged vertebrate lies less with laboratory prowess than with evolutionary history. The question we should ask is not whether it is possible to have a six-legged vertebrate, but why land-living vertebrates should have precisely *four* legs, arranged in two pairs, rather than some other number of arrangement of legs. Why *shouldn't* vertebrates have six legs – or, indeed, any other number of legs? Just because all the land-living vertebrates we see happen to have four legs, this does not mean that no other arrangement is possible.

Here, by way of elaboration, is a related analogy. Until the late 1980s, all anatomists assumed that the hands and feet of land-living vertebrates always had five digits – the equivalents of the thumb, index, middle, ring and pinkie fingers. To be sure, many animals had fewer than five digits per limb, but such cases could be seen as reductions from the original, basic number five. In the late 1980s, however, paleontologists began to unearth ancient fossils of the earliest amphibians — very primitive, potentially land-living vertebrates — and made the surprising discovery that the number of digits was variable. The salamander-like creature *Acanthostega*, for example, had eight digits per limb.[4] It seems that very early on in the evolution of digits, evolution experimented with different numbers, and the number five happened to be the one decided, ultimately, by history and chance. This discovery shattered the preconceived illusion that the number five was somehow the manifest destiny of digit number, and that it was possible to imagine a world in which the usual number of digits was eight, or twelve, or three, or some other number.

Could the same contingency be true for limbs, too? The limbs of land vertebrates evolved from the fins of large, shallow-water river fishes with two pairs of fins — one fore, one aft — available for evolving into limbs. With the real-world example of the evolution of digits as a guide, it is possible to imagine an alternative course of evolution in which six-legged vertebrates evolved from fishes with three or more pairs of fins, rather than two. However, genetic and paleontological evidence suggests that the tendency for a fore-and-aft arrangement of two pairs of limbs goes very deeply in vertebrate evolution,[5] perhaps back to the days when primitively jawless and limbless fishes, related to the modern lamprey,[6] were first evolving fins. Despite an early variation in the number of digits, it appears that having two pairs of limbs is something truly fundamental to the character of vertebrates.[7]

One way out of this seeming box canyon is to hark back to homeotic mutations and suggest that, yes, although vertebrates have definite fore-and hind limbs, each with its own identity, there need be no restriction on the number of *pairs* of fore or hind limbs. In other words, it might be possible to have a single pair of hind-limbs, yet two pairs of forelimbs (one of which evolved into wings), making six limbs altogether. In other words, what is fundamental about vertebrates is not the number of limbs *per se*, but the number and character of *limb fields* from which limbs might develop. This kind of variation — in the numbers of limbs of certain kinds that appear in animals — is very prominent in the evolution of insects but has not been observed in vertebrates, and there is no evidence that vertebrates ever experienced this kind of limb duplication. But as one of the oldest scientific sayings goes, absence of evidence is not evidence of absence.

11. THE EYES OF LEGOLAS GREENLEAF

When asked to describe the physical nature of the Elves in human terms, Tolkien would suggest that they were like humans of heroic stature, additionally blessed with freedom from disease and the cares of age, and gifted with various powers that humans would find distinctly alien. The inhuman, incomprehensible nature of Elvish sleep is variously mentioned, as is their lack of fear of the dead, given that the Elves simultaneously live in the worlds of dead and living. It is also clear that Elves have far more acute powers of perception than mere human beings. There is a famous passage in *The Two Towers* in which Legolas and Aragorn, gazing over the plains of Rohan, see riders approaching. Where Aragorn — who is, in the estimation of no less a person than Gandalf, the greatest huntsman and tracker of the age (*Rings* I,2) — sees just a swiftly moving blur, Legolas can distinguish 105 horsemen and discern their hair color and the points of their spears and even differences in their height, from a distance of five leagues — and has the temerity to make light of this feat (*Rings* III,2).

When summarizing the superhuman capacities of the Elves, it would be tempting to say that Elves can see further than humans, but this would be inaccurate: Aragorn can see just as far as Legolas, but he cannot resolve the blur he sees into individual points. What is at issue, then, is not distance, but something called 'acuity' or 'resolution' — the ability to resolve objects as distinct, from a given distance.

Thinking about this further (and doing a few quick calculations) I realized how Tolkien's apparently throw-away remarks, made to illustrate the physical capabilities of the Elves, could have had a profound effect on the way Elves would have seen and reacted to the world.

What do I mean by 'distance' and 'resolution'? Well, consider the Andromeda Galaxy, the closest major galaxy to our own Milky Way. Anyone with normal vision who lives in the Northern Hemisphere can make it out on a clear night as a fuzzy patch of gray, even though it is at least 2.5 million light years away — the most distant object that can be seen with the unaided human eye.

However, it is not possible to resolve any of the hundreds of millions of stars within the Andromeda Galaxy as individual objects without an enormous telescope. This instrument wouldn't bring the Andromeda Galaxy any closer: all it would do is increase resolution, that is, the degree to which you might see closely spaced objects (stars, in this case) as individual points, rather than a hazy blur of all the stars together. The power of telescopes is rated according to their resolution, not by how far they can 'see.'

Resolution is measured as angular distance 'of arc,' that is, in degrees (°), minutes (′) and seconds (″). The Celestial Sphere is 360 degrees all the way round — a perfect circle. Each degree is divided into sixty minutes, each minute into sixty seconds. Two objects in the sky are said to be separated by an angular distance measured in degrees (or minutes, or seconds) of arc.

So just how acute are the eyes of the Elves, and how would this affect the way they saw their world? Let's start with the episode mentioned above, which offers plenty of clues. Legolas says that the riders are hardly more than five leagues away from where he and Aragorn are standing. Let us assume that they are *precisely* five leagues away — that is, fifteen miles — and also that a rider on his horse, viewed from the front, is about a yard across, and that the riders are at least a yard apart.

Some simple trigonometry on the back of an envelope shows that to resolve a single rider at that distance — that is, to resolve two points a yard apart at a distance of fifteen miles — Legolas' keen eyes would have to have an acuity of

$$2 \times \tan^{-1} (0.5 / 26400)$$

where 26400 is the number of yards in 15 miles (15 x 1760), which works out at about two thousandths of a degree (0.002°), or just under eight seconds (8˝) of arc. This means that Legolas would be able to separate two points separated by just two thousandths of a degree. We can take this as a benchmark, although given that Legolas was hardly straining himself — he could see the color of the hair on the riders' heads, spear points, and even detect that the riders varied in height — he would have had greater acuity even than this, although a reference earlier in the same chapter suggests that he has trouble distinguishing individuals in a company of Orcs some twelve leagues away. Nevertheless, Legolas is described as being keen-sighted even for an Elf: on the final march to the Black Gate (*Rings* V,10), only Legolas can actually see the high-flying Nazgûl, so we can take eight seconds of arc as a benchmark for the acuity of the average Elf. How would this compare with the limits of human visual acuity?

Traditionally, the limits of human visual acuity are tested with a pair of stars, *Mizar* and *Alcor*, a double-star system in the constellation of the Big Dipper. The Arabs, who gave the stars the names we conventionally use, called these stars the 'horse and rider.' It is possible — just — for a keen-sighted person on a clear night to resolve *Alcor* and *Mizar* as two separate stars of unequal brightness (*Alcor* being the smaller and dimmer of the pair), rather than as a single star.

Alcor and *Mizar* are separated by 12 minutes of arc, so this can be said to stand as the limit of human naked-eye resolution. Objects closer together than this will only ever be seen as a blur, no matter how keen-sighted the (human) observer. When you turn your telescope on the problem, it turns out that *Mizar* (as distinct from *Alcor*) is itself a double-star. The two stars are separated by 14 seconds of arc, well within the abilities of Legolas, if not Aragorn.

To complicate the picture further, each of these stars is its itself a double, making *Mizar* a *quadruple*-star system. The stars in the brighter pair of the two doubles are separated by seven or eight thousandths of a second of arc, requiring a thousand times the acuity that Legolas showed on the fields of Rohan, a feat that might have defeated even him. However, the acuity of the Elves already approaches what is possible given the blurring effect that the Earth's atmosphere has on stars. Viewed from space, stars are pointlike — when viewed from the Earth's surface beneath a soup of air, stars are smudged out into discs approximately one second of arc across, so resolving the four stars of *Mizar* would not be possible even for a phenomenally keen-sighted Elf unless he were launched into space.

However, this comparison between the abilities of Men and Elves, even fully earthbound ones, gives us an impression of how the world was seen through the eyes of the Elves. Aragorn at his most acute would only ever have seen *Alcor* and *Mizar* as a double-star, but Legolas would easily have seen it as a naked-eye *triple* star (*Alcor*, and the stellar pair that make up *Mizar*), an object unknown to human sight. The starry sky was of great significance to the Elves, and no wonder: they would have seen far more in it than human beings ever could.

The Elvish night sky would have been far more lively than the human one, as well as being more detailed, because greater acuity means an increased ability to detect movement. Planetary motion provides one example. The planets were first described, by the ancients, as the stars that 'wandered' against the background of the 'fixed' stars.

This wandering could be charted by plotting this movement, and it was shown that the planets all describe a full circle around the sky, although at varying speeds, complicated by the fact that we ourselves live on a moving planet (which makes it look like some of the planets appear to reverse direction now and then, but we needn't worry about that here.)

The planet Mars, for example, takes 1.8822 years to travel the full 360 degrees, and leaving out all other complications for the sake of simplicity, it therefore moves an angular distance of 12 minutes — equivalent to the limits of human acuity — in just under nine hours. This means that it would take an unaided human observer nine hours to notice that Mars had moved at all, with respect to the background of the sky. Given that Mars is probably not observable for nine hours together on any given night (because of the Earth's rotation, apart from such hazards as cloud and moonlight), an unaided human observer would not be able to detect the motion of Mars against the background of stars in any one session.

Elves wouldn't have to wait nearly that long. Taking eight seconds of arc as something an Elf such as Legolas was comfortable with, he would have noticed the motion of Mars in just six minutes of observation, or less. To an Elf, the Sun, Moon, planets, and other phenomena such as shooting stars would have fairly whizzed across the firmament.

This argument doesn't just apply to stars. When Galadriel sadly sings of the leaves of Lórien blowing in the wind (*Rings* II,8), she would have been conscious of a host of tremulous movement invisible to human eyes, making the trees seem even more alive, and giving her lament added poignancy. The eyes of the Elves are keen indeed, with a sharpness that pierces the heart.

The tremendous acuity of Elvish eyes raises the question of how they physically managed it. In one sense, acuity is all about information transfer: that is, how fast the eyes can gather information from the environment and process it to create an image. The reason that powerful telescopes tend to be large is that the need to be able to devote a large area (such as a big lens, or a mirror) to the job of collecting the stray photons of light emanating from a distant source, such as a faint galaxy careering towards the edge of the observable universe. If all light sources emanate (or reflect) light spherically, that is, equally from all over their surfaces, it is easy to see that proportionately less light emanating from a distant source will be traveling in our direction than from a source nearby. This is why distant sources appear fainter than ones close by.

The way to compensate for faintness is to focus on the object of interest for a long time, allowing the telescope to collect as much light as convenient for the formation of a usefully bright image. This is why telescopes often take many minutes, or even hours, to take pictures of distant celestial objects, and why even amateur astronomers will equip their telescopes with motor drives that allow their telescopes to keep pointing at the same object, compensating for the movement of the Earth that would otherwise constantly drag the object from the field of view.

Photographers will be familiar with this problem: when photographing objects that are faint, distant, or both, it is essential to maximize the exposure time: that is, how much time the aperture of the camera can be left open, minimizing blur. This is all very well for taking pictures of a static object, even in low light, but can be quite a challenge for capturing fast-moving events, in which keeping the shutter open for minutes or hours is not possible. The photographer may be a long way from the split-second, close-up moment that he wishes to capture. This means that shutter speeds may be kept very short (a thousandth of a second, or even less) otherwise the picture will be blurred. In such a situation, a camera will have to work hard to capture as much light as it can in as short a time as possible. This requires lenses that are extremely acute, like Elvish eyes — that is, they can resolve closely spaced objects from a long way off.

In photographic terms, the acutest lenses are those with the longest *focal length*. This is a property determined by how the lens is shaped, and is a measure of the distance between the lens and the point at which rays of light collected by a lens converge on a focus. For general-purpose, everyday shots, cameras typically have a focal length of 35-50mm. For close-ups, or distant shots, a focal length of 90-200mm might be needed, and sports photographers use lenses with even greater focal lengths, 500-1000mm or more. A small telescope suitable for studying the Moon would typically have a focal length of this order. Long-focus lenses fill the same frame as shorter ones, but do so by blowing up the tiny detail of interest. Long focus, therefore, implies greater angular resolution, and therefore greater acuity.

In a sense, Legolas's eyes had greater focal length than Aragorn's, as he was able to pick up the riders of Rohan from a great distance and 'blow up' the image to fill his mental picture in more detail than Aragorn's wide-angle view.

The downside is that the greater the focal length and the 'narrower' the view, the harder it is to collect enough light to make a photograph: an expression of the same physical law that governs why more distant objects are fainter than closer ones. Making lenses more acute, then, implies a loss of light-gathering capacity — unless, of course, the lenses are made larger to compensate. This is why the lenses of sports photographers are physically large, combining long focal length with the big, broad lenses needed to gather as much light as possible in the fractions of a second available.

In photography, the light-gathering capacity of a lens (as opposed to its acuity) is rated in terms of its *focal ratio*, or *f*-ratio.

This is the focal length of the lens divided by its diameter. The smaller the *f*-ratio, the greater the diameter of the lens will be in relation to its focal length, and so its light-collection ability will be better than a lens of smaller diameter but the same focal length. To put it another way, the lens of a given focal length that has an *f*-ratio of 4, say, will have four times the area — the essential measure of light-gathering capacity — than one with the same focal length but rated at *f*8, and this is reflected in its size and price.

However, the *f*-ratio of many camera lenses is not fixed, but can be adjusted by varying the amount by which the shutter opens, by means of a device called an 'iris.' The iris can be adjusted to cover more or less of the lens, altering its effective area. For example, a lens rated at *f*4 can be 'stopped down' to *f*8, say, or *f*16. (Note how the stopping goes down, even though the numerical value goes up.) This might seem perverse, given all I've said about the need to gather as much light as possible in a short time, a situation in which lenses should always have the smallest focal ratios possible.

However, there is more to focal ratio than gathering light. Every photographer knows that a consequence of varying the focal ratio is a concomitant variation in what is called the *focal depth* — the degree to which the focus can vary before the image blurs. It so happens (indeed, it's a simple consequence of geometry) that it is much harder to keep a lens focused on an object if the lens has a low *f*-ratio than if it has a high one. That is, low *f*-ratio implies a narrow depth of focus. Photographers use this effect if they want to emphasize a particular feature of the scene: when taking a picture of that sports moment, or that illicit kiss, a low *f*-ratio means that the key moment is kept in focus, but everything else is blurred. Contrast this with the landscapes of the famous photographer Ansel Adams (1902-1984), whose frequent use of stopped-down (that is, high *f*-ratio) lenses meant that *everything* in the scene was in focus, from the stones in the foreground to the clouds and mountains behind.

It so happens that the eyes of humans and other animals have irises, too, and they work very much in the same way as the iris in a camera, varying the focal ratio of the eye, both to modulate the amount of light coming in, and to vary the depth of focus. So, after this long digression into optics, we can explore how the remarkable visual capacities of the Elves might be tied into anatomy.

The iris of the human eye controls the width of the pupil — the dark center of the eye that is, in physical terms, its light-gathering aperture. The iris moves automatically as conditions vary, a process known as *accommodation*. If we want to focus on a particular object, our pupils dilate, or widen, effectively decreasing the focal ratio of the eyes, narrowing their depth of field and blurring anything in which we are not interested. We never see the world in the all-over focus of an Ansel Adams landscape (explaining why his pictures have a somewhat haunting, unearthly quality.) At the same time, our pupils dilate when the ambient light is low, allowing us to gather as much light as possible, at a cost of some blurriness in the general field of view. In bright light, however, our pupils contract, cutting down the amount of light that enters the eye, but broadening the depth of field.

Human beings are used to using both eyes at once, which allows for the easy estimation of depth or distance by a process of triangulation. However, accommodation means that it is possible to judge distance very finely with just one eye, by opening up the pupils fully and restricting the depth of focus to the object of interest. Chameleons, for example, have eyes that move independently, and their ability to shoot their tongues out at flies with deadly accuracy shows that they can judge distance very finely, even with only one eye. This is achieved by opening up the pupil as widely as possible — even in very bright light — narrowing the depth of field to focus on the prey, and nothing else.[1]

This lesson on optics makes the eyes of the Elves even more remarkable. Legolas's visual acuity demands very long focal length, and yet the ability to take in extremely fine detail at a glance, that is, with a very quick shutter. However, his eyes are not the telescopic lenses of the sports photographer: as far as we know, they are no different from that of a human in its basic dimensions. To compensate for lack of focal length when viewing distant objects, therefore, Legolas must have had very large pupils indeed, permitting formidable light-gathering ability combined with a very fine depth of focus that would have allowed him to judge distances very accurately as well as get very detailed images in no more than a glance. To do this, Elvish pupils would have had

122

to have filled virtually the entire visible surface of the eye, leaving very little iris or exposed whites. This extreme degree of accommodation, however, might not have been sufficient for the telescopic ability of Elvish eyes, so some compensation might have been achieved within the brain itself, coming back to my original theme – that the brains and minds of Elves could do strange things, barely comprehensible to a mere human.

Everything I have said so far about cameras and optics would be true for old-fashioned systems in which the image is formed by the reaction of light on silver salts contained within a photographic emulsion. The sensitivity of a film to light can be tuned by varying the size of the mineral grains in the emulsion. In general, the coarser the grains, the more sensitive the overall film to light, and the quicker it can be exposed, at a cost of the degree of resolution. 'Fast' film can be used in low light, or with rapid shutter speeds, but tends to give grainier results than 'slow' film. This graininess is easily seen in blow-ups of newspaper sports shots. However, this is where the hitherto close analogy between cameras and human (or Elvish) eyes breaks down, as the way we form images is very much more involved than simply the passive smearing of light over a photographic film.

Once light has passed through the lens of the eye, modulated by the pupil, it strikes the retina, a layer of light-sensitive cells at the back of the eye. Just as it does in a photographic emulsion, the light impinging on the cells of the retina triggers a chemical reaction. From then on, the analogy breaks down. The light-stimulated chemical reactions in the retina do not simply stay there, but set in train a series of nervous impulses that make their way, in stages, to the visual centers of the brain. During this process the image is manipulated and transformed in various ways, so that what we *think* we see is not necessarily representative of the sum of raw photonic energy that strikes our eyeballs. At the simplest level, our picture of the world is turned the right way up (because the image that hits the retina is, actually, upside-down.) At the most complex, our image of the world is massaged and edited by our expectations, habits, associations, previous experiences and even interactions with other senses.

In a way, a truer analogy of the process of vision is not with a traditional camera but with a digital imaging system coupled to a photographic editing program, in which the image, once received, can be manipulated to emphasize, distort, or even remove certain features. Even very simple digital snapshot cameras will have inbuilt software to allow 'digital zoom' — focusing on particular parts of a scene in a way that could not be achieved by the camera's optics. In this way, the computer becomes very much part of the camera, presenting an image that is pleasant, or at least immediately useful to the needs of the viewer, rather than necessarily or objectively truthful.

One might imagine that the visual superiority of the Elves had less to do with any physical difference between Elvish and human eyes, than to do with the way that the image, once received, was processed. It is reasonable to suggest that Elvish nervous systems are much more refined than their human equivalents, with greater sensitivity and power. If that is the case, then the Elvish visual system was far more efficient than the human one, able to extract more meaningful information from a given stream of photons, and — if the information is lacking — interpolating the lost 'pixels' to complete a scene. When Legolas surveys the riders of Rohan, his eyes are, in effect, being augmented by a sophisticated digital zoom system. It is his brain, rather than his eyes, that 'zooms in' on the riders of Rohan, excluding any other, possibly distracting features of the landscape.

It occurs to me that some of the features of Elvish eyesight could be explained by their nocturnal origins. The legends of the Elves suggest that they 'awoke' in Middle-earth before the Valar had created the Sun and Moon, and that they are naturally equipped for life under the stars. Men, however, 'awoke' later, once the Sun and Moon had been established (*The Silmarillion*, chapters 11, 12), and would have been adapted for life in stronger illumination. It could be that some of the acuity of the Elves might be explained by their adaptation to a nocturnal existence, so that they could make the most of very low light levels by means of mechanisms of light-collection and visual processing far more efficient than found in Men, who are adapted to abundant sunshine and moonlight.

This idea immediately seems to run into problems. We know from our own experience that it is very hard to see things in any detail in very low light, partly as a consequence of accommodation. We also know that color is very difficult to discern in low light, and that we are normally color-blind at night, the world fading into shades of gray. This problem, however, is an illusion, caused by our own limitations as daylight-adapted creatures. This limitation does not appear to have affected the Elves (or if it did, it is not mentioned) whose lives are filled with color, even under starlight. Our own inability to see color at night, for example, does not reflect physical reality, but results from the way our eyes are made — night-flying moths, for example, whose eyes are made very differently from ours, can see in color in very dim starlight, conditions in which humans are completely color-blind.[2]

Elves, though, did not have the multifaceted eyes of moths. As far as we can judge, their eyes looked like those of humans for all practical purposes, so how could they have seen color at night? Our inability is a consequence of the finite abilities of the cells in the retina that respond to light. These are divisible into two classes, called 'rods' and 'cones' by virtue of their shapes. Rods are highly efficient detectors of light of all visible frequencies (that is, colors), but make no distinction between one color or another. Cones, on the other hand, tend to be tuned to narrower frequency bands: our retinas contain three types of cone cell, maximally responsive to red, green, or blue light, but at the cost of some efficiency. Rods function well in conditions of low light in which cone cells hardly work at all, explaining why we are color-blind at night.

Even in daylight, our ability to see certain colors varies from person to person. My very first car, for example, was a fetching shade of green — or so it seemed to me, even though everyone else I knew said that it was in fact blue, a contention supported by its description on its official registration documents. It could be that I have more green-than blue-sensitive cone cells: it is known that the proportions of the different types of cone cell vary between people, sometimes markedly. There is a strong genetic component to color sensitivity, and some of the genes responsible are carried on the sex chromosomes. This explains why the incidence of the common forms of hereditary color-blindness is strongly determined by sex, in that these syndromes are almost entirely restricted to males.

There is also some research to show that some females may have more than three types of cone cell, in that there are two varieties of the red-sensitive cone; that it is possible in females to have both at once; and that such gifted individuals have a richer perception of color than is the case with people generally.[3] Perhaps fancifully, this could explain the frustration of women unable to get their men folk to express opinions about the color of clothing or interior furnishings: and why it is that small girls, but not small boys, are fascinated by the color pink. Animals with color vision may have more or fewer varieties of cone cell than three, and their perception of color will be respectively more or less refined than in humans.

The point of this is to suggest that if the types, numbers, and capabilities of color-sensitive cells varies among humans, and between humans and animals, then it is possible to imagine that the Elvish retina contained rod and cone cells that might have been different from those found in human ones; that they were more sensitive to light, and in particular that Elvish cone cells could function well in twilight, which is not the case in humans.

That Elvish eyes were, in the beginning, adapted for greater sensitivity to light is proved in the breach by references to creatures initially adapted for daylight but had had darkness thrust upon them. I refer, of course, to Gollum, in origin a Hobbit-like creature whose long residence in complete darkness had enlarged his eyes and had even made them luminous — as if they had evolved a 'tapetum,' the reflective, light-enhancing structure seen in the eyes of night-hunting carnivores such as cats. That such structures could appear in a single lifetime of one creature is absurd, and reflects the fundamentally Lamarckian (as opposed to Darwinian) mode of evolution in Middle-earth.

The Elves, of course, were characteristically having their cake and eating it too. Creatures adapted for life in twilight often have great difficulty coping with brightness. In *The Island of the Colorblind*, Oliver Sacks looks at the unusually high incidence of hereditary color-blindness in the isolated atolls of Micronesia, and shows that as well as the inability to see color, people without cone cells do not have the visual acuity necessary to read well, and also suffer greatly from the brightness of the tropical sun. Perhaps in mitigation, one might say that although the Elves were creatures of starlight, they would have experienced brightness in other ways, such as by exposure to the Two Trees: perhaps, then, the Elves were adapted for a greater *range* of brightness than that experienced by humans.

Now that I've established the concept that different creatures might perceive color differently, according to the properties and disposition of light-sensitive cells, it is easy to imagine cases in which creatures might be able to see things that are *invisible* to human eyes. After all, the light-sensitive cells in human eyes are sensitive to light in a relatively narrow range, and some creatures might be able to detect light outside this range.

What we think of as visible light represents just one small window in a continuum of photon-mediated radiation called the 'electromagnetic spectrum,' of which the familiar rainbow spectrum of visible light is a part. The spectrum, in both its wide and narrow sense, is organized by the amount of energy carried by the photons of radiation, which is related to the 'frequency' of the photons — the amount by which the waves associated with the photons oscillate in a given interval of time — and their 'wavelength' — the physical distance between adjacent crests of these oscillations. Radiation of the highest energy has the shortest wavelengths and the highest frequencies.

'Visible' light is of intermediate wavelength and frequency, ranging from red light with a wavelength of about 650 nm (nanometers, where a nanometer is a millionth of a millimeter) to violet light at 400 nm. Beyond the violet, the light becomes the ultraviolet (or UV), which is invisible to our eyes, because we do not have receptors tuned to see it. Still more distant, with progressively higher frequencies, are X-rays, g-rays (gamma rays), and cosmic rays. These forms of radiation carry more energy per photon than visible light, and have a greater ability to penetrate matter such as, for example, human skin: the UV rays in sunlight damage our skin, causing tanning – a reaction to UV-induced damage. X-rays, however, are more powerful still, and can penetrate the entire human body. High-energy gamma and cosmic rays, typically products of titanic explosions elsewhere in the Universe, are very dangerous indeed and can cause serious damage to people and equipment. Space is awash with this 'hard' radiation, but we on Earth are protected, to a large extent, by the atmosphere.

Just beyond the red end of the visible spectrum is the 'infrared' (or 'IR'). This radiation is less energetic than visible radiation, carried by photons of lower frequency and longer wavelength. We perceive infrared radiation as heat: the red glow of a fires represent emission that strays into the visible range. Beyond the infrared is the extensive part of the electromagnetic spectrum inhabited by radio waves, reporting the lives of stars and galaxies, the most distant echoes of the birth of the Universe, as well as TV pictures and radio news from closer to home. The wavelengths of radio waves are measurable in meters, even kilometers, but are nowadays more usually referred to in terms of their frequency. Radio signals in the FM waveband will be carried by radio waves with a frequency of around 100 megahertz or MHz (million cycles per second).

Animals cannot 'see' extremes of radiation. For example, there are no cases of animals that detect X-rays, because the photons are too energetic to be safely trapped. Similarly, there are no animals that detect radio waves, because the radiation is too faint and diffuse (a fact that explains why radio telescopes with acuities comparable with that of the telescopes discussed earlier in this chapter are physically very much larger.) Nonetheless, the range of radiation detectable by organisms is considerably wider than the 'visible' range of between 400 and 650nm. As I have discussed, our bodies can detect and respond to some UV and infrared radiation, even if we do not use it to construct images. However, the regions of UV and IR immediately adjacent to the visible range are widely exploited by the sensory systems of organisms. IR cameras that detect body heat are well-known, and some animals, notably snakes, are able to use heat sensors to track prey — although these organs have nothing to do with eyes.

The petals of some flowers, in contrast, have patterns and markings visible only to bees, which can see ultraviolet light: these patterns presumably have colors which are simply indescribable in human terms except in the intellectual language of frequency and wavelength. It is also known that birds of prey such as the kestrel (*Falco tinnunculus*) hunts its rodent prey by the use of UV cues completely invisible to humans. The voles (*Microtus agrestis*) hunted by the kestrels live a life dominated by smell, and mark their trails through vegetation with urine and feces. Unbeknownst to the voles, their waste products are strong absorbers of UV light: because of this, vole trails are highly visible to kestrels flying above, presumably as dark streaks against a lighter background — just as if they were following roads. Moreover, the strength of the UV absorption provides a good measure of whether the vole trail is busy or not, allowing the kestrel to judge whether it is worth stooping to take a closer look.[4]

There is strong evidence that Elves (and possibly Dwarves) are able to see things outside the human visible frequency range. This comes from the several references to various kinds of invisible writing such as 'Moon Letters,' or the *ithildin* letters on the Gates of Moria, visible only to humans (and Hobbits) under highly proscriptive circumstances, such as the presence of heat (in the case of the fiery letters on the Ring); the position of the Moon, or the light of the setting Sun on the last day of autumn, and so on. However, it is possible that Elves were able to see such things all the time, or perhaps more often than humans. I refer in particular to a passage in *The Fellowship of the Ring* (*Rings* I,12) in which the Elf Glorfindel, meeting the travelers on the road, examines the hilts of the Morgul-blade that had been used to stab Frodo on Weathertop.

Glorfindel notes the presence of inscriptions on the blade that had been completely invisible to Aragorn. Now, it could be that the letters were written or engraved in extremely fine or small script, but it is more likely, in my view, that they were written in normal-sized writing that was visible to Elvish eyes only, because they were strong absorbers of UV but not visible light. This incident illustrates the remarkable abilities of the Elves very well, even more than the episode on the plains of Rohan. Where Legolas simply has better visual acuity than Aragorn, Glorfindel is capable of seeing into a visual realm that is completely beyond Aragorn's experience. Back on the fields of Rohan, at the moment when Aragorn reveals his lineage and majesty before the astonished Éomer, only Legolas can see a flickering crown of white flame on Aragorn's brow.

12. OF MITHRIL

At the beginning of the Second Age, Dwarves migrated from their ancient cities in the Blue Mountains to swell their mansions at Khazad-Dûm, or Moria. Many Elves went with them and founded Eregion, or Hollin, a small kingdom on the western side of the Misty Mountains (*Rings* II,4). Elves and Dwarves alike had been drawn by the rumor of *mithril*: a metal harder than steel, as lustrous and as ductile as silver, incomparably rare and of a value beyond estimation: Gandalf says that the corselet of *mithril* rings gifted to Bilbo by Thorin at the end of *The Hobbit* was probably worth more than the entire contents of the Shire (*Rings* II,4). In *The Lord of the Rings* as published (*Rings* II,4), *mithril* is described as uniquely a product of Moria, although drafts describe it as having been found in very small quantities elsewhere (*HOME* VII), including the Ered Luin (*HOME* XII). Nevertheless, it was the discovery of substantial veins of *mithril* beneath Caradhras in the Misty Mountains led to the spectacular rise of Moria, and its subsequent fall, after the search for *mithril* ever deeper beneath the mountains awoke a sleeping Balrog.

Although much of the physical fabric of Middle-earth — its rocks and minerals — is recognizable and prosaic, *mithril* stands alone as something special, not comparable with anything in the real world. On the level of the story, *mithril* is as much a fantasy creation of *The Lord of the Rings* as Elves and Hobbits, and as such does not require a specific identification with anything in the modern world. In any case, it is a tacit assumption that the last accessible veins were exhausted by the end of the Third Age, so if it were once a real substance, there isn't any more of it: or if there is, it would be too difficult to find.

OF MITHRIL

The attraction of *mithril* lies in its inaccessibility. In what follows, therefore, I am not going to say that *mithril* 'really was' this substance or that. To do that would destroy the magic, and, in any case, I do not believe that Tolkien ever left any specific notes about its identity. However, that should not prevent us from looking for materials in the modern world that might have the properties of *mithril* that Tolkien did describe, if only in the spirit of a riddle game.

Let us begin with what Tolkien told us about marvelous *mithril*. It was a shiny metal somewhat like silver in appearance, easy to extract, lighter yet stronger and harder than steel, and yet very ductile, which means that it is easily worked by a smith or jeweler at room temperature, or at worst with the help of the kind of forge available to any blacksmith. The density of *mithril* is, however, ambiguous. In drafts of the scenes in Moria (*HOME* VI, VII) Gandalf says that 'Moria-silver' is almost as heavy as lead, but this idea was presumably abandoned, as it would make wearing a mail shirt of *mithril* rings much more onerous for Frodo than it evidently was. Although when wrapped up as a parcel, the *mithril*-coat seemed heavy for its size, Bilbo told Frodo that it felt much lighter when worn (*Rings* II,3): however, armor of any material feels much lighter when you put it on rather than when you have to carry it by hand, because its weight is evenly distributed over the body. When forced to wear an Orc mail-coat in the Tower of Cirith Ungol (*Rings* VI,2), Frodo comments that it is heavier than this (now lost) *mithril* mail.

Finding a single material that is both strong and ductile at relatively low temperatures is a tall order, because these properties seem to be contradictory. Strong metals have a microscopic crystalline structure that resists deformation until a critical point is reached, when they start to crack. This means that strength often goes with brittleness, which is precisely the wrong property of a metal that is easily workable. Can you have a metal that is strong and ductile at the same time, or will *mithril* forever remain in the realm of fantasy?

Let us break down the problem and see what kinds of materials might suit. *Mithril* is obviously a metal, and metals come in three varieties: they may be chemical elements; they may be alloys in which various elements are mixed together to suit the purposes of the engineer; or they are chemical compounds, that is, substances in which the constituent chemical elements are combined in fixed ratios. Water, for example, is a compound in which there are always two atoms of the element hydrogen to every one of oxygen — H_2O.

Most of the metals we are familiar with are chemical elements. This means that they are pure substances that cannot be broken down into anything simpler. None, however, manage to be both strong and ductile at the same time. If an elemental metal is ductile, it is also weak. This tends to apply to those metals that are most easily extracted from their parent rocks, and therefore known to metallurgists since antiquity: elements such as gold, silver, copper, mercury, tin and lead. If an elemental metal is strong, such as iron, it is likely to be heavy, brittle, or both. Lightweight, lustrous metals that might be suitable candidates for *mithril* — metals such as the magnesium, aluminum, or titanium of today's high-tech applications — must first be extracted from their ores using prodigious amounts of energy and the kind of heavy, industrial-scale equipment more associated with the forces of darkness than with the Dwarves, and certainly not the Elves, who would have preferred more elegant and less obviously 'smokestack' means of manufacture (see the chapter entitled 'Indistinguishable from Magic' for more about the philosophy behind Elvish technology.) However, we know that *mithril* was easy to extract, and occurred in identifiable veins, so it need not have been one of these hard-to-extract metals anyway. *Mithril* as a chemical element seems to be out. 'True silver' remains plain old silver, nothing more.

Alloys might offer an easy alternative. Elemental iron on its own is hard, but it rusts. Add a little vanadium, chromium, and carbon, though, and it becomes an alloy called stainless steel, which is rustproof and much harder than pure iron. Brass is a mixture of copper and zinc, and bronze a mixture of copper and tin. There is even an alloy called electrum, a combination of silver and gold in various proportions used in ancient times for coinage.

Although alloys are invariably man-made, one could imagine *mithril* as a natural alloy of, say, iron, silver, and a little carbon. The alloy might also incorporate the extremely light elementary metal beryllium, given the fondness of the Elves for green semi-precious minerals such as beryl ($Be_3Al_2Si_6O_{18}$), a variety of aquamarine.

However, it seems to me that Tolkien's *mithril* always has a distinct identity, and properties recognizable as such down the ages. This would not be true of an alloy, whose properties can change according to the recipe used by the metallurgist — it would also mean that the *mithril* found in Moria need not have the same chemical constituents as that found anywhere else.

If *mithril* is neither a chemical element nor an alloy, the only choice left is that it is a chemical compound, a substance made of fixed, unvarying ratios of chemical elements, chemically bound together rather than simply mixed. It so happens that there is a class of chemical compounds called 'intermetallics.' Like alloys, intermetallics consist of two or more metals combined. Unlike alloys — and this is crucial — the metals are always in fixed ratios, because intermetallics are true chemical compounds, as opposed to mixtures or alloys. The intermetallic compound Ni_3Al, for example, is not an arbitrary mixture of nickel and aluminum, but a compound with a definite crystal structure in which there are always three nickel atoms for every one of aluminum. Intermetallics, like alloys, are products of technology rather than nature, but is it possible to think of *mithril* as a kind of naturally occurring intermetallic?

Well, nearly. Intermetallics seem to have many of the properties of *mithril*. They are durable, strong, and shiny, and they have all kinds of interesting magnetic and electrical properties that might be useful should you wish to invent, say, *ithildin*, the *mithril*-based substance used for the inscription on the doors of Moria, visible in starlight and moonlight (*Rings* II,4). But there is a huge downside: intermetallics are very brittle. This deficiency of chemical character has prevented the widespread use of intermetallics in technology and industry, and would be no good for *mithril*, either.

So we seem to be stuck. Or we would be, were it not for the extremely recent discovery of a family of simple intermetallics that are shiny, strong — and ductile.[1] They all consist of a regular metal, such as copper or silver, allied with one of a member of the intriguing 'rare-earth' metals. This group of substances includes such exotic elements as cerium, scandium, and lanthanum, as well as four elements named after a single place. These elements are yttrium, ytterbium, terbium, and erbium, all of which were first extracted (and named after) one extraordinarily productive quarry at Ytterby, near Stockholm, Sweden — along with their fellow rare-earths holmium (from *Holmia*, the Latin name for the city of Stockholm), thulium (from 'Thule,' the Latin for the furthest north), and gadolinium (after one Johan Gadolin (1760-1852), the Finnish chemist who discovered yttria, the ore from which yttrium was first extracted.)

Rare-earths have all kinds of intriguing properties. Lanthanum, and to a greater extent yttrium, have come to prominence recently as they are essential ingredients in one of several novel classes of substance capable of conducting electricity without resistance at far higher temperatures than previously attainable. Before the mid-1980s, this phenomenon, called 'superconductivity,' was only observable at temperatures a few degrees above Absolute Zero (minus 273°C). However, the discovery of superconductivity at higher temperatures than this in substances such as yttrium barium copper oxide caused a sensation, and led to Nobel Prizes for Georg Bednorz and Karl Alex Muller, the scientists at IBM in Switzerland, who discovered a formulation of lanthanum barium copper oxide that superconducted at minus 238°C, twelve degrees above the previous record-holder.[2] A flurry of research soon followed. The current well-established record for superconductivity is minus 140°C in a material comprised of mercury, barium, calcium, copper and oxygen (but, sadly, no rare-earths).[3] However, the promise of superconductivity at room temperature has not yet been attained, and the mechanism of superconductivity remains imperfectly understood.

Away from the headlines, many rare-earth elements are ductile, which is why researchers have been exploring their use as ingredients for the long-sought intermetallic compound that would be workable, and thus commercially useful. One of the more promising candidates is yttrium silver, an intermetallic in which atoms of silver and yttrium occur in precisely equal amounts. Yttrium silver is so ductile that a wire can be stretched to a fifth again its length before it snaps. There is something about its crystal structure, not yet fully understood, that allows it a degree of plastic flow without its breaking into ragged fragments.

The ductility of yttrium silver, along with its luster and as-yet-unexplored electronic and magnetic properties, would make it a good real-world candidate for *mithril*. Yttrium silver is, however, totally synthetic. As far as we know, no seam of the stuff waits to be mined out of some unexplored cavern. On the other hand, exotic minerals exist that are known from single localities on Earth, formed by circumstances of temperature and pressure unique to that place only. Who is to say that the silver in the Silverlode did not contain a little yttrium?

One property of *mithril* I have not discussed is its 'hardness.' In Peter Jackson's film of *The Fellowship of the Ring*, Bilbo compares the hardness of *mithril* with that of dragon scales. Allowing that Bilbo was being poetic or figurative rather than scientifically accurate, it is a safe bet that dragon scales — when not additionally encrusted with gems as the result of protracted repose on a Dwarvish hoard — are coated with enamel, the hardest substance produced by living creatures, and the same substance that coats your teeth. So how hard is that?

When seeking to identify a gem, jewelers assess hardness by the ability of one substance to scratch another. This concept was formalized in 1822 by the German mineralogist Friedrich Moh, and the Moh scale of hardness is still used today. In the Moh scale, diamond has a hardness of 10. Diamond is still the hardest known naturally occurring substance — nothing matches it, or scratches it. Corundum (that is, ruby or sapphire) can be scratched by diamond, and has a hardness of 9. Topaz (with a hardness of 8) can be scratched by rubies but itself scratches quartz (hardness of 7), and so on, all the way down to flaky gypsum (2) and powdery talc at the bottom, with a hardness of 1. On this scale, teeth come in at around 5, the hardness of the mineral apatite – which is not surprising, as this is the same mineral that forms the basis of bones and teeth. One would expect *mithril* to be at least as hard as that: good-quality steel has a hardness of around 6.5, so Bilbo's equation of *mithril* with dragon scales is about right.

Mithril is just the right thing to make a tough corselet of mail rings. But any skilled smith might have enhanced its properties still further, by controlling the crystalline structure of the material – an important determinant of its hardness. Swordsmiths can control the properties of blades by forging them in just the right way to ensure that the edge is extremely hard and brittle, but the main body of the blade is flexible. A key step in the forging of a samurai sword involves wrapping the sword in clay, with only the edge exposed. The sword is then heated to above what is known as the 'martensitic' transition, when the steel adopts a different kind of crystalline configuration. The hot blade is then thrust very suddenly into a pool of cold, stagnant water (Elven swordsmiths can be seen doing just this during the reforging of the sword Narsil in Peter Jackson's film of *The Return of the King*.) The rapid cooling of the edge produces a fine-grained microcrystalline structure that is very hard, but very brittle. The bulk of the blade, shielded by the clay, cools more gently, resulting in a structure with much larger crystal grains, fewer defects (which tend to inhabit the areas between grains), and therefore greater resistance to fracture.

Although a mail-shirt of *mithril* rings isn't quite the same thing as a sword, one might imagine how a smith might put it through a quench-hardening process. This would leave the core of each ring strong but flexible, and the outer surface brittle but very, very hard — and with a network of microscopically fine cracks that might give the material a sparkly appearance if turned in the light.

Quench-hardening does not work with metals that do not undergo the martensitic transition, but the elastic properties of metals more generally can be controlled by varying their micro-structure through careful manufacture. There is, however, an intriguing alloy of nickel and titanium that undergoes the marten-sitic transition — and which, therefore, could be quench-hard-ened. The wonderful property of this alloy is that it is a 'memory' metal — it is light and flexible, but springs back to its original shape once bent. These properties make it ideal for high-tech spectacle frames, even though it was originally developed for use in the light-weight, super-tough protective shells of the warheads of intercontinental ballistic missiles, designed to stop them burn-ing up on atmospheric re-entry.

Mithril has many wonderful properties, but how could a thin, flexible mail-shirt of any material have resisted the sudden, con-centrated impact of a spear wielded by an Orc chieftain, so that Frodo — under the full impact of the spear — could have survived with only a few bruises, much to the staggerment of the Fellow-ship (*Rings* II,5)? To answer this question I have had the benefit of the advice of Eichling, who wrote to me care of TheOneRing.net to share more than twenty years of experience of using and abusing plate and mail armor made from various materials.

She has recently been armored in a hauberk made of titanium chain mail, the closest she could get to *mithril* for lightness, and which is reasonably good at resisting mild point impacts. Although small arms were not available to Tolkien's protagonists, Eichling found that titanium chain mail when draped over a log stops a 9mm round fired from 10 feet. One ring from the mail was lost, but the bullet failed to penetrate the log beneath, suggesting that the titanium mail absorbed the impact. The problem with titanium is its lack of mass. A heavy steel hauberk would probably offer more protection. The ambiguity about the mass of *mithril* – heavy when held in the hand, but light when made into a mail-shirt – makes it hard to judge how *mithril* would have reacted to point impacts. Nevertheless, the fact that Frodo's mail-shirt resisted the impact of a spear wielded by a troll suggests that something else is at work besides the property of the metal itself. The secret could lie in how the shirt is made rather than what it is made from.

Going back to first principles, Frodo's shirt needs to be made in such a way that it is loose and flexible in normal use, but responds, by virtue of its weave, to a sudden impact by stiffening immediately into a solid shell, distributing the force of the blow evenly all over the structure. This sounds like an absolutely magical property of armor – as magical as an invisible ring – and Eichling's titanium hauberk does not behave this way. If it did, the 9mm shell would have bounced off, leaving the mail shirt undamaged. There are, nevertheless, real-world materials that behave just like this, and at least one of them can now be found in armor.

One of these materials has been found in a toy-box near you for more than half a century. That material is called 'Silly Putty.' The story of 'Silly Putty' began in 1943 when one James Wright, an engineer working for General Electric, was working out how to make synthetic rubber. He added boric acid to a silicone oil, producing a very bouncy putty-like substance for which no practical use could be found. Marketed as a toy in 1950, 'Silly Putty' has become an evergreen plaything. In addition to its remarkable bounce, 'Silly Putty' has the curious property that it is very pliable and stretchy when mild force is applied, but when hit with a hammer it will resist the impact and keep its shape. This behavior is a consequence of its structure as a viscous 'gel' – that is, a substance made of long, chain-like molecules distributed in a liquid medium, the changing properties of the substance determined by the ways in which the molecules interact with one another.

Decades further on, work on advanced materials has produced materials such as TF2, a synthetic, foam-like substance found in the protective clothing worn by motorcyclists. Loose and pliable during normal wear, TF2 is like 'Silly Putty' in that it becomes rigid when subject to sudden acceleration and impact. This is just the way that Frodo's *mithril* coat appears to behave: however, it is clear that this mail coat is as other mail coats, made of interlinked metal rings, not some other, exotic substance. It could be that the weave of *mithril* threads or rings in Frodo's mail shirt behaves as a macroscopic version of the chemical structure of 'Silly Putty,' so that it is flexible in normal wear, but becomes very stiff when hit suddenly by an Orc-wielded spear.

13. THE LABORATORY OF FËANOR

In this chapter I delve into the fascinating world of Elvish technology. At first, the terms 'Elvish' and 'technology' seem uneasy bedfellows, but as I explained in the Introduction, we should not be blinded to what appears to be Tolkien's aversion to science. After all, the core of Tolkien's legendarium concerns the adventures of the Noldor, cited specifically as that caste of the Elves most interested in the acquisition of knowledge, its organization, and its practical use (see, for example, *Letters* 183) and how jealousy over the fruits of that knowledge led to disaster. Despite Gandalf's anti-reductionist tirade to Saruman (*Rings* II,2), it was not the knowledge that was evil nor the hunger for it, but the hoarding of it, or its use in the acquisition of power or domination over others.

Elvish science reached its peak in the First Age with Fëanor, universally regarded as the brightest and most powerful of all the Elves.[1] It was Fëanor who created the Silmarils, the great jewels over which the first wars against Morgoth were fought. The *palantíri* or 'Seeing Stones' were gifts from the Elves to the Faithful of Numenor, having been made long before in Valinor by the Noldor, possibly even by Fëanor himself (*Rings* III,9). It was Celebrimbor of Hollin, a descendant of Fëanor, who forged the Three Rings. Many important incidents in *The Silmarillion* and *The Lord of the Rings* turn on the fruits of the technology of the Noldor. In this chapter I discuss two of these — the Silmarils and the *palantíri*. The One Ring, forged by Sauron, poses special problems, and so gets a chapter on its own.

The Lord of the Rings contains many passing references to the relative hardness of materials, but this hardness has a mythic quality in that it directly correlates with technological sophistication of the smiths associated with that substance. For example, the Ents easily destroy the country rock that forms the outbuildings and walls of Isengard, but they are unable to make a dent in Orthanc, a tower built by the long-vanished Numenoreans, a tower that Gandalf says cannot be destroyed from without (*Rings* III,9; III,10).[2] But when Wormtongue tosses the *palantír* of Orthanc from an upstairs window, it makes a distinct chip in the Numenorean step on which it falls — a step against which the rage of Treebeard has had no effect at all. As regards hardness, the *palantíri*, made by the Elves of the Ancient West, are to Orthanc what Orthanc is to regular building stone, and this is reflected in the durability of the materials whence they are made.[3]

So what is the extremely durable substance from which the *palantíri* are made? First of all, let us examine some of the other remarkable things that the *palantíri* can do, and ask whether real-world materials exist that share these properties.

Tolkien writes (in *Unfinished Tales*) that the *palantíri* permit the sight of small images of things far away. As described, then, they are no more than telescopes, but given their properties as described in *The Lord of the Rings* they are more like 'hypertelescopes,' able to image things not in a direct line of sight, or obscured by intervening matter, such as clouds or mountains. But the main purpose of *palantíri* is to act as communicators. Leaving the psychical aspects of *palantíri* aside, it is possible to imagine two *palantíri* being in communication by a phenomenon called 'quantum entanglement.'

Quantum entanglement is a process in which two particles (such as photons of light) are generated together in such a way that they are, in the quantum sense, aspects of the 'same' particle: being intimately entwined, they can potentially remain in communication with each other when the particles are far apart: so that if the quantum state of one of the particles is altered, this alteration will show up in the other particle, no matter where in the universe it is. Nothing shows up the inherent weirdness of the quantum world so much as quantum entanglement — what Einstein called 'spooky action at a distance' — yet the phenomenon is already being explored in such fields as advanced computing and cryptography.[4] If Fëanor had made all the *palantíri* at once so that they shared the same quantum 'state,' they would still be able to communicate with one another despite being separated. A change in the environment of one of them — such as the impression of the thoughts of a person in the vicinity — should be communicated to the others, instantly.

How could this be done? The tale lies in the making. One of the most intriguing products to come out of research into advanced materials is lithium niobate, a glassy substance with a blue-green cast that has all kinds of interesting optical properties. One of these is called 'parametric down-conversion:' shine a single photon of light at a piece of lithium niobate, and two photons will emerge. These photons will have twice the wavelength of the original (that is, their color will have shifted to the red end of the spectrum), but the key thing is that they will be entangled. Lithium niobate looks like a very promising ingredient for a *palantír*.

But there is more to seeing stones than communication in some abstract sense — they must also be able to communicate images. Lithium niobate is promising here, too: if you shine a bright enough light on it, the electrons in the material can be persuaded to shuffle around, changing the optical properties of the material. If matters are arranged in just the right way, you can create a hologram: basically, an image can be made to appear inside the material, as if it were a crystal ball.

However, lithium niobate, for all its wonderful properties, is unlikely to be as hard as *palantíri* so obviously were, so some other substance is required. Happily, I have just the thing, and it is called beta carbon nitride. This fabulous substance, it has to be said, is almost as mythical as the *palantíri* themselves. Calculations suggest that it should be even harder than diamond — that is, its hardness would go to 11. The trouble is that nobody has managed to synthesize it, although some scientists believe that they might have come close.[5] Alpha carbon nitride, on the other hand, is already known as a grayish substance, and stands to beta carbon nitride as the graphite in your pencil stands to diamond, graphite and diamond both being forms of the same substance, in this case pure carbon.

I propose that the *palantíri* were made of thousands of alternating, microscopically thin layers of lithium niobate and beta carbon nitride. This layered structure would have preserved the hardness of the substance but would also have allowed a little flexibility, so that a *palantír* would have bounced if thrown from a high window – rather than shattering, as a diamond might. Such a layered construction would have given the *palantíri* a pearl-like iridescence, caused by the interference of light with the alternating layers: pearls are iridescent precisely because they have this kind of structure, and this in turn suggests a wonderfully Elvish mode for the manufacture of *palantíri*. Rather than having been carved, forged, or molded, the *palantíri* would have been *grown*, layer by layer, in a similar way to the growth of a pearl by the accumulation of layers of nacre. When Fëanor made the *palantíri*, they were his characteristically Noldorin answer to the pearls that his Telerin cousins scattered across the beaches of Eldamar.

While Fëanor was creating the *palantíri*, he might have put beta carbon nitride and lithium niobate to another use. Tolkien explicitly states (*The Silmarillion*, chapter 7) that the substance of the Silmarils will never be known until the world is unmade: a proscription which I view, naturally, as a challenge. The Silmarils could have been made from alternating layers of lithium niobate and beta carbon nitride, or perhaps a core of lithium niobate surrounded by a hard shell of beta carbon nitride.

You will recall that Fëanor used the Silmarils to capture the immortal light of the Two Trees of Valinor; that they appeared to shine of their own radiance, and that they enriched and amplified any light that fell on them. Without wishing to speculate on the nature of the Two Trees themselves,[6] their images could have been preserved holographically within the Silmarils, exploiting the properties of lithium niobate. But the Silmarils could also have exploited yet another property of lithium niobate, rather like parametric down-conversion, but in reverse. That property is called 'second-harmonic generation.' This means that in certain rather contrived conditions the material halves the wavelength of light that is shone on it, shifting it towards the bluer end of the spectrum. In effect, this would allow a Silmaril to absorb non-visible radiation, such as heat, and re-transmit it as visible light, giving the impression that the Silmarils would shine of their own accord — especially in the presence of heat-producing living tissue, whether the hand of Beren or the crowned head of Morgoth, as well as responding to ambient light by shining even more brightly.

14. THE GATES OF MINAS TIRITH

There is a scene in *The Lord of the Rings* in which Legolas the Elf and Gimli the Dwarf walk through the streets of Minas Tirith, the Capital City of Man, musing on the human condition (*Rings* V,9). Gimli notes the impermanence of human endeavor: how the early promise of human activity is almost invariably stunted, or goes awry. Nevertheless, says Legolas, Man will still be here long after Elves and Dwarves have faded from view. This, for me, is perhaps the most poignant scene that Tolkien ever wrote, for it points up what I believe to be the most important theme in Tolkien's work — *loss*.

Loss was a theme of Tolkien's life, both personal and professional, and a preoccupation with loss inevitably found its way into his fiction. As noted in Humphrey Carpenter's *Biography*, Tolkien's early life was marred by the loss of his father and mother; the loss to urbanization of the rural west-Midland haunts of his boyhood; the loss of any sense of a home as the circumstances of his life became progressively more makeshift and nomadic; his enforced separation from his *fiancée*; and the loss of all but one of his immediate circle of friends in the Great War. The loss of most of his family as a result of his mother's conversion to Roman Catholicism instilled within him a fierce and defiant love of his adopted religion, as well as reinforcing his inherent pessimism.

Against this background, it is hardly surprising that Tolkien, bereft and largely homeless, became, like Sméagol, a wanderer obsessed with roots and beginnings, and that he chose, as his field of study, the very darkest of the Dark Ages, from which only the most fragmentary and enigmatic literary remains survive. In his professional work, Tolkien was very like a paleontologist obliged to make do with the tiny amount of evidence that posterity has given us, and was awed and intrigued by the vast gulfs of the unknown that surround what few precious scraps of illumination we have. As has been pointed out with reference to *Beowulf*, the central text of the Early Middle Ages in England, Tolkien's importance lies in his insistence that the poem must not be studied as an artifact in isolation, but as an example of a tradition of storytelling whose context is now lost, but nevertheless must have been there for the poem to make sense.[1]

When we look at Tolkien's fiction, loss is everywhere, and in various guises. Right from the start, Tolkien wished to recreate, through his fiction, the lost mythology of pre-Christian England which he felt must have been there *in extenso*, but now discernible only from tiny hints in place-names and obscure references in poems such as *Beowulf* and more explicitly Christian works such as the *Crist* of Cynewulf. Turning to the themes of the fiction itself, *The Lord of the Rings* is a quest not to gain a prize but to lose one. But the loss of the One Ring and the diminution of the powers of darkness is also accompanied by the extinction of much that is good: it is the final knell for the Elves, whose tenure in Middle-earth throughout the Third Age has been artificially prolonged by the existence of the Three Rings, bound to the One — as if the preservative effects of the Ring on Bilbo or Gollum were expanded to a whole culture.

For the Elves, loss is tinged with yearning and regret, a motif that goes all the way back to the loss of the Straight Road at the end of the Second Age; the loss of Beleriand and the Silmarils at the end of the First; the loss of the Two Trees even earlier, and even the marring of Arda at the beginning of days. The Elves, being effectively immortal, can remember these wounds as if they were made yesterday — wounds which, to the countless generations of mortals, have faded into the custom of myth and fairy-tale, their once high significance worn to the conventional mindlessness of nursery verse.

This kind of loss — this literary loss, if you like — is just the kind that Tolkien addressed as a professional philologist, linguist and excavator of myth. His reconstructions of lost languages such as Gothic builds on it, as does his entire view of *Beowulf* and more fragmentary narratives. It is this kind of loss from which our own lore has accumulated, and Tolkien provides some startlingly concrete examples. Perhaps the best known is the nursery-rhyme that begins *Hey diddle diddle*:

> Hey diddle diddle, the cat and the fiddle,
> The cow jumped over the moon;
> The little dog laughed to see such fun,
> And the dish ran away with the spoon.

At first sight this is plain nonsense, and could have been made up from any old string of syllables recited, for example, to calm a fractious infant. In the *Oxford Dictionary of Nursery Rhymes*, Iona and Peter Opie note that all that can be said for certain about the origin of this best-known of nursery rhymes is that it appeared in print around 1765, and that all the many hypotheses to explain the origin of the verse are very likely as nonsensical as its content. If so, this would give Tolkien a completely clear run to invent whatever derivation he chose.

Therefore, asks Tolkien, what if *Hey Diddle Diddle* is the threadbare remnant of a once more sensible story, transmitted orally from generation to generation, losing lines at each transmission, until it degrades finally into nonsense hardly more informative than noise? Tolkien gives his own answer in a poem, recited by Frodo at *The Prancing Pony* (Rings I,9). In the poem, the Man in the Moon takes a break from his celestial course, tethers his luminous charge to a hilltop, and rolls down to sample the famously good beer of a welcoming inn (just the kind of thing that Tolkien might have done himself). Apart from its excellent beer, the inn has an assortment of bizarre inhabitants including a tipsy fiddle-playing cat, a frisky cow, a dog that appreciates jokes, animated tableware and so on. The Man in the Moon outstays his welcome and, being too drunk to move by himself, has to be helped back into the Moon by the regulars before the Sun gets up. Although whimsical in the extreme, the poem does at least have a narrative that is barely suggested in the nursery rhyme that survives.

This theme of literary loss, in which old tales of heroism get worn down into Old Wives' Tales and nursery rhymes, crops up repeatedly in the *Lord of the Rings*. To take just one example, when Éomer confronts Legolas, Gimli, and Aragorn on the plains of Rohan (*Rings* III,2), he expresses astonishment at the sudden apparition of beings from old fairy tales out of the North – ironic in view of the close literary links between the Rohirrim and *Beowulf*: doubly ironic in that the entire conceit of Middle-earth is couched as Tolkien's reconstruction of English myth-lore from *Beowulf* and such few scraps we have of pre-Christian English mythology.

To take a purely scientific (as opposed to literary) parallel, the slow erosion of information into repetitive idiocy resonates with what we know of the evolution of the human genome, in which time and chance have made nonsensical drifts of huge swaths of once useful genetic information, the accumulations so mountainous that they constrain the culture, as it were, of the relatively few functioning genes that remain. By way of analogy, our own culture is built on a deep layer of lore, fragmented by time, but in its dissolution and rot providing a rich loam on which our present culture is built.

Perhaps closer to our present concerns is the loss not of culture, but of technology. There are several instances in Tolkien's work where the maker of a precious but lost artifact complains that they no longer have the ability to create anything to match it. The entire plot of *The Silmarillion*, in fact, turns on this very point: after the Two Trees of Valinor are destroyed by Morgoth, the goddess Yavanna, who grew them to begin with, complains that she no longer has capacity to replace them from scratch. The only chance of salvation lies in breaking the Silmarils, in whose structure Fëanor had trapped the light of the Trees before their destruction. Fëanor, for his part, complains that were the Silmarils to be broken, he would also lack the capacity to make anything as fair again (*The Silmarillion*, chapter 9).

Fëanor's refusal to surrender the Silmarils is crucial, for it sparks the rebellion of the Noldor and the entire adventure in Middle-earth without which the events in *The Lord of the Rings* would not, ultimately, have happened. This relationship between makers and artifacts is echoed in the story of the Elven Rings and their makers, and, indeed, between Sauron and the One Ring itself:

once these great artifacts are made, their makers lose the capacity to repeat the feat.

This phenomenon can be seen to have a spiritual rather than a technological dimension. However, hints of explicitly technological loss can be found elsewhere. For example, Glóin mentions to Frodo (*Rings* II,1) that any works that the Dwarves might undertake now — that is, at the end of the Third Age — would not match the craftsmanship of their ancestors. Outside the doors of Moria, Gimli comments that the craft of making such doors had long vanished from the world, along with their makers (*Rings* II,4). At the Houses of Healing (*Rings* V,8), we are told that the 'leechcraft' — that is, medical knowledge — of Gondor was less than it had once been. Elsewhere, Gandalf tells Frodo that Saruman has long sought the lost secrets of the manufacture of magic rings (*Rings* I,2) – so deeply lost, apparently, that even the Wise among the Noldor, such as Galadriel and Elrond, could not simply have told him how. The various characters in *The Lord of the Rings* marvel at the artifacts of ancient Gondor — the Pillars of the Argonath, the Tower of Orthanc, and so on — comparing them with the diminished and decayed state of Gondor in their own time.

The word *orthanc* is an intentional pun: in Sindarin it means 'Mount Fang' but in Old English it means cunning smithcraft. Lines 405 and 406 of Beowulf read:

> Beowulf maðelode   on him byrne scan,
> searonet seowed   smiþes orþancum

Which translates as

> Beowulf spake, his breastplate gleamed,
> war-net woven by wit of the smith[2]

These lines suggest that the craft – the 'wit,' the *orþancum* — is somehow special, arcane, or occult, and, in the sense of Saruman's tower, the product of a technology now lost, and moreover, forgotten. For most of the characters in *The Lord of the Rings*, the construction of artifacts such as the Argonath and Orthanc defies comprehension to the extent that these things may just as well be the products of natural forces as of human craftsmanship, in rather the same way that the Anglo-Saxon invaders of Britain tended to view the monuments of the vanished Romans as the work of *ettens*. Like the tales and legends whose context is lost by the passage of time and, once broken free of the moorings of significance, degenerate into doggerel suitable only for burbling to babies, the characters at the end of the Third Age view such artifacts simply as part of the scenery: as the landscape which, unbeknownst to them, has shaped their lives, their characters, and their existence.

This aspect of Middle-earth as an element that shapes the story by means of its topography is emphasized in certain passages in which the landscape itself comes to the fore and plays a more active part in the narrative. One thinks, for example, of the episode in which the Old Forest connives to direct the hapless Hobbits towards fates they did not intend (*Rings* I,6), and that in *The Hobbit* where the travelers, caught in a storm in the Misty Mountains, witness 'stone giants' throwing boulders at one another. Before one wonders what part these giants, never encountered anywhere else, play in the bestiary of Middle-earth, one might consider that they are there not as discrete entities but as personifications of the thunder-riven landscape, rocks and stones made animate. Tolkien, in fact, drew this scene from life, recounting many years later an Alpine walking holiday in which the party was bombarded by snowmelt-loosened boulders, caroming from far above (*Letters* 232).

Such explicit personification strikes us as unusual today, but it would have been less so to Tolkien, who was forever drawing on literary conventions of an earlier age. The audience of *Beowulf*, for example, would have been familiar with figures of speech such as *hringiren scir song in searwum* — 'bright iron rings sang on byrnies'[3] — in which the qualities and actions of living things are poetically ascribed to inanimate entities. In other words, the singing of the armor is literal as well as metaphorical. Taking this idea to its logical extreme, one might imagine that Tom Bombadil and Goldberry — always problematic as characters on the level of the Hobbits, or even the Elves — are no more than personifications of aspects of the landscape, a possibility admitted as such by Tolkien himself (*Letters* 19, 153).

This aspect of Tolkien's work finds a parallel with the fiction of Thomas Hardy. One commentator has noted how the landscape of Hardy's Wessex dwarfs its human inhabitants, whose fates are in part determined by its topography.[4] The concept of characters helplessly at the mercy of their environment, but, by virtue of the brevity of their own lives unable to comprehend why, is very close to the idea of fate, destiny, or *wyrd* with which Old English poets and audiences were familiar.

Technology plays an important part in the slow degeneracy seen in Tolkien's tales. As far as they are concerned, the characters of *The Lord of the Rings* are living in a world in which swords, spears, bows, and arrows have been the stock in trade of warriors since time immemorial. This impression of technological stasis is, however, erroneous. We have already seen how the Elves and Dwarves rue the loss of craftsmanship available to their ancestors.

The same is true for Men, but far less explicitly, and the reason is that by the end of the Third Age, the technology of Man has descended proportionately far more than that of Elves or Dwarves, to the extent that Men living at the time of the story are unaware of the depths to which they have sunk. Consider: the near-unbreakable stones of Orthanc are of Numenorean — that is, *human* construction, not Elvish, far surpassing the skills of Third-Age Gondorians. And there are hints of even greater marvels, buried under strata of cultural amnesia so deep that these wonders, created by *smibes orbancum*, can hardly be described in terms of the language accessible to people at the end of the Third Age. For example, after the Downfall of Numenor when the world was bent, Tolkien obscurely suggests (*The Drowning of Anadûnê*, in *HOME* IX) that some Numenoreans built ships which, they hoped, would travel in genuinely straight lines through the air, leaving the curve of the Earth far below. It is as if the Ancient Minoans once built airplanes, but all the records we have are myths about the dangers of flying too close to the Sun while wearing wings made of feathers and wax.

And in case we are inclined to say that we, in our modern scientific world, have taken these lessons to heart, such a conclusion would be no more than hubris. We still marvel at how the ancients built monuments such as Stonehenge and the Great Pyramid. And closer to our own time, we can only gasp in awe at how astronauts were once sent to the Moon in spaceships no more technologically sophisticated than washing machines — and that despite our nominally greater technological sophistication, no Man has set foot on the Moon since 1972. Unless we pick up the threads and resume manned space exploration, the Apollo adventure of thirty years ago will quickly fade from history into legend, and finally into myth neither more nor less compelling than that of Daedalus and Icarus.

Perhaps most pertinent to this book is Tolkien's near-universal preoccupation with a kind of loss that is environmental and even anthropological. *The Lord of the Rings* takes place in a world that is curiously empty, even though it is clear from the context (and the density of ruined fortresses, barrows, and so on) that the population was greater, once upon a time. Whole countries — Eregion, Enedwaith, Minhiriath, the Brown Lands — have been laid waste by war and disease and never repopulated. Centuries of strife have left the once-populous kingdom of Arnor almost completely desolate – all, that is, apart from the Shire, whose inhabitants are almost completely unaware of the destruction all around them. Gondor, the great Kingdom of Men that once dominated the Westlands, has been reduced by slow attrition to a city-state comprising Minas Tirith and the lands round about.

Those few inhabitants that remain in this depauperate wilderness are themselves largely on the long, downward slope to extinction. I have already discussed the extinction of the Ents, but the point is made repeatedly and relentlessly that the Elves are also on their way out, just enough remaining to maintain havens and way-stations for a general migration away from Middle-earth (in which case their extinction is a mixture of removal and fading rather than mortal finality, but the effect on the constitution of Middle-earth is much the same.)

Of Men, much is made of Aragorn being the last of an etiolated pedigree. The Dwarves, too, survive on a diet of cultural nostalgia, and the legions of cannon-fodder created by Sauron are maintained artificially. Only Hobbits seem to be thriving, but this picture of prosperous and unchanging stability might be interpreted as immediate stagnation against a trend of very slow decline. The Hobbits show hints of a history in which they were once more widespread – the colony of Hobbits in Bree, for example, and the very existence of Gollum, the last survivor of the now-extinct Hobbit-like River Folk of the Vales of Anduin.

The slow and steady diminution as recorded is entirely deliberate, for it seeks to explain how the non-human peoples depicted with great realism in *The Lord of the Rings* have become, in the present day, residents exclusively of fairy-tales in which they are found only as grotesque caricatures of their former selves. The Elves, a woodland-dwelling people of beauty, power, and — in characters such as Galadriel — awe and terror, diminish to the fey and filmy sprites of woodland glades. The Dwarves, doughty residents of the Mountain-halls, become — via the brothers Grimm – the knockabout characters of Disney's *Snow White and the Seven Dwarves*, a film released in 1937, the same year as *The Hobbit* .[5]

It is perhaps significant in this context that Tolkien's original vernacular rendering of the name 'Noldor' — the most powerful and scientifically gifted of all the Elvish kindred — was 'Gnomes,' a word which was derived from the Greek root gnw-, associated with words implying thought or knowledge (whence 'gnomic,' 'gnomon' and so on), but in the modern world more associated with the kitsch sculptures that adorn suburban gardens.[6] How has the shade of the mighty Fëanor, greatest of the Gnomes, been laid low!

As a process, the diminution of Elves and Dwarves (not to mention Orcs) from heroic characters to the bottom-of-the-garden fairies of nursery-rhyme is exactly the same as the slide of literature from coherent heroic verse into half-forgotten snatches of doggerel. It is this process of fragmentation that Tolkien found so centrally absorbing — both as a student of a literature that had almost wholly vanished but for the chance preservation of the barest scraps of which *Beowulf* is the best example, and as a creator of the lost landscape of Middle-earth. It is a fascination shared by many scientists, especially but not exclusively paleontologists and archaeologists, who must reconstruct the past from fragments bequeathed not by careful memorialists but by the mindless circumstances of fate.

The atmosphere of loss in Tolkien's work, especially in *The Lord of the Rings*, is all-pervasive and at times almost unbearable, and I find it remarkable that people have not made more of it. Shippey comes close in *J.R.R. Tolkien: Author of the Century*, in his suggestion that the overriding theme in Tolkien's work is death, but even then he is wide of the mark. Death, as a theme in Tolkien, is parochial, being a special instance of loss. In any case, the theme of death does not encapsulate the slow, generations-long process of erosion that precedes a final extinction that may never, in fact, take place. The Elves do not die as long as the Earth endures – they just diminish and fade: we still *have* garden gnomes, *Snow White* and *Hey Diddle Diddle*, even though the larger and grander stories whence they came have been forgotten. The contrast between the mortality of Men with the immortality of the Elves is less a homily about the fact of death, than on the torture of having to remain tied to the Earth for long enough to witness (and, therefore, to mourn) processes of decay too slow to be noticeable to mortals, whose impression of the world is, at worst, one of stasis. The gift of death, it is said, is something that with the passing ages even the Gods will grow to envy.

Critics schooled in the minutiae of modern literature rather than the sweep of epics – still less, the chilly magnitudes of science — fail to appreciate this dimension of loss and, as a consequence, tend to take Tolkien's Elves and Dwarves at face value, comparing them *directly* with the their bedisney'd and degenerate remnants, with no understanding of the slow process of figurative diminution that links the one with the other. They are then left wondering why a distinguished Professor should have made so much of what, to them, is so much infantile fantasy. As I hope I have explained, such critics will have missed the point.

The poignancy of the conversation between Legolas and Gimli is informed, therefore, by a *double* sense of loss – first, that the Elf and the Dwarf know perfectly well what the *wyrd* of the world has in store for them, and, second, that Men are in general ignorant of these tides of time and fate that will leave them as sole inheritors of the Earth. This poignancy, furthermore, resonates with what recent research has uncovered about the actual history of Man — research that is fascinating because it stands in opposition to the conventional mythology about our own lost past.

It is customary to think of human evolution as a process in which there is an inherent drive towards improvement, in which species replace one another in an orderly way, as if they were baton-carrying runners in a relay race. This linear idea is, however, a caricature of what really happens, as I explained earlier on in my discussions on the evolution of Orcs. It has become more and more apparent that human evolution, like the evolution of any other species, is quite unlike a ladder and more like a bush, in which many collateral lines of human-like species evolved together, with the implication that many different species of human coexisted on the Earth.[7] Indeed, such coexistence would have been the usual state of affairs, from the appearance of demonstrably bipedal forms around four and a half million years ago to the disappearance of Neanderthal Man less than 30,000 years ago. To have one, single species of human in possession of the whole planet is extremely unusual, occupying less than one per cent of the entire history of the human lineage as a distinct entity.

As recently as 100,000 years ago, at least four species of human existed on Earth, when the first fully modern members of our own species *Homo sapiens* emerged from Africa, its ancestral home. At that time, Europe and western Asia was inhabited by Neanderthal Man, *Homo neanderthalensis*, and it is possible that the late, last remnants of the extremely ancient species *Homo erectus* still survived in Indonesia. Other, poorly known species, distant cousins of Neanderthals, lived in China and other parts of eastern Asia.

Neanderthal Man appeared around 300,000 years ago and the last material records of this species date to less than 30,000 years ago. More controversially, it has been argued that *Homo erectus*, which appeared around two million years ago, hung on in Indonesia until as late as 27,000 years ago.[8] These estimates are based on scarce material remains, and all paleontologists recognize that the latest occurrence of a species in the fossil record need not represent the very last member of that species to have existed. In all probability, Neanderthalers and Java Men would have persisted to yet more recent times.

In my book *In Search of Deep Time*, I spun just such a scenario as a way of trying to imagine how our species, *Homo sapiens*, might have related to these other species, in many ways so human: in many others, so completely different. I imagined Neanderthalers surviving as picturesque mountain villagers in medieval Spain, the last examples slaughtered as demons by Ferdinand and Isabella in 1492. I imagined two small jungle populations of naked, mute *Homo erectus* discovered in the twenty-first century. One, in Sumatra, was immediately wiped out by loggers, and the depredations of hunters accustomed to stealing infant orang-utans for the exotic pet trade. The other, in the remote and inaccessible mountains of Annam on the Laos-Vietnam border, would survive slightly longer, but even there, habitat pressure would force their evacuation. The last example, an aged male, died in a zoo in Hanoi in 2014, aged 26.

In the context of Middle-earth, this story serves to make two points. The first is to highlight the fact that Men, in Middle-earth as in the real world, remain largely unaware that for most of their history they were accompanied by other, similar species — in Tolkien's terms, other 'speaking peoples.' Tolkien makes the point that the Elves and Dwarves of fairy tale need not have been idle fancies, but the natural end-products of a long process of decay leading to disappearance, first in body, and then in memory. Even if they exist no longer, Neanderthals and *Homo erectus* were once very real, far more so than folk-tales of yetis and the *orang pendek* of Malay folklore might suggest.

The second point is more general, and concerns the poignancy of any loss in the diversity of species and cultures available to us. I do not mean to say that the Elves and Dwarves of our stories represent folk-memories of the Neanderthals and *Homo erectus* with whom we once shared our world, than to comment on a richness and diversity of species, culture, and viewpoints that the world once held, that is now gone. Elves and Dwarves have their own idiosyncratic cultures and perspectives on the world, informed by differing biological perspectives and enriched by millennia of tradition – all of it soon to disappear, with Men hardly being aware that things that were once great and marvelous have vanished altogether. What is most touching and, ultimately, tragic about the conversation between Legolas and Gimli is that is *they*, the creatures soon to disappear, are remarking on that fact.

The message is also relevant to us, today. Not only is the world dominated for the first time by just one species of human, the diversity of culture and language even in this single species is being eroded rapidly. Languages are vanishing as their native speakers grow old and die out, their children having learned some more dominant language. This phenomenon is active from the forests of Africa to the Arctic wastes of the Inuit. Only with immense efforts can such threatened languages be preserved. In the British Isles, the indigenous Celtic languages of Cornwall and the Isle of Man have ceased to exist as viable vehicles of discourse. Irish, Welsh, and Gaelic survive thanks to concentrated intervention, bulwarks against the ubiquity of Modern English. When a language becomes extinct, the cultural heart is torn from a people, because it is in distinctive languages that peoples best preserve their cultural identity, hero-myths, legends, and folk tales.

Tolkien was more sensitive than perhaps any other author to the catastrophic potential of linguistic extinction. He spent his professional life breathing literary life into the few remaining fragments of the ancient, lost tongues of the north — Old Norse, Gothic, and Old English, languages in which were expressed ideals of the virtues of courage and heroism that seem so alien and inexplicable to speakers of modern English, our universal argot conditioned by what Tolkien thought of as the insult of Norman French, among many other things. His need for restitution was so great that he was driven, from his teenage years, to create new, lost languages — and then, of course, he had to invent a culture and a mythology to go with them.

15. THE LIVES OF THE ELVES

Why do we grow old and die? Why is our allotted span the Three-Score Years and Ten of the Bible, and not much longer, or much less? And however long we live, why do human beings not vary greatly in the rates at which they age? To be sure, some people look old before they are thirty, whereas others are still hale into their nineties: but why aren't there people who measure out their lives in minutes — or millennia?

Why do different species age at different rates? The tiny laboratory roundworm *Caenorhabditis elegans* — a creature much used by researchers interested in the genetics for aging — never lives much beyond three weeks. Most of the fauna you see in your garden in summer will not make it through the winter. Your pet hamster will, very likely, show signs of extreme age at two years. Your dog could see a good fifteen years before senile decrepitude sets in. The cat, however, might see twenty or more, and yet it will die before a human being has grown to full maturity. Few warm-blooded creatures can match human beings for longevity. Whales, parrots, and elephants are among the few. The real old-timers are cold-blooded creatures such as turtles and some fishes, along with ancient fungi and certain coniferous trees such as sequoias and bristlecone pines that can live for thousands of years.

A human being lives far longer than one might expect for an animal of its size. But can we — *will* we — ever be relieved of the necessity of death? And were we offered such a gift, would we embrace it greedily, or would we, like Faramir and the Ring, shun it even were it to be found as a trinket abandoned on the highway?

Although Tom Shippey remarks (in *J.R.R. Tolkien: Author of the Century*) on the importance of death as a theme in *The Lord of the Rings*, few have investigated the biological consequences of Elvish immortality for the lives that Elves must lead, irrespective of the contrast that is inevitably drawn between this immortality and the mortality of Man.

Tolkien did, however, identify with our preoccupation with the kinds of questions I ask above. As our technological power and mastery over the Earth increases, so does our obsession with aging and death. We do things that we know will shorten our lives — smoking, drinking alcohol, using drugs, eating too much, working too hard, suffering the stress that only headlong ambition and relentless, largely futile activity can confer – while simultaneously buying, in huge quantities, potions and elixirs of all kinds on the promise that they might extend life, but whose efficacy, as we are perfectly aware, is dubious. How similar we are to the Numenoreans at the height of their power, Tolkien seems to say: in command of the Earth and yet obsessed with death, clinging to life until we fall witless from our chairs, unable to halt the shortening of the once centuries-long lifespan enjoyed by our wiser and more respectful ancestors. Scientists in the real world are beginning to understand why we grow old and die, and why it is that different species tend to age at different rates.

However, nobody really knows why the human lifespan seems limited to just over a century, or if there is anything that can be done to extend it significantly. In contrast, the sentient residents of Middle-earth — the 'speaking peoples' — enjoy a remarkably various range of life-spans.

Nothing is said explicitly of the aging of creatures such as Orcs, though it might be assumed that the more 'manufactured' kinds would have had a rather limited lifespan. Each sprung fully-formed into adulthood, much like a butterfly from a chrysalis, it is likely they would have lived only as long as necessary to perform specific tasks. If they did not die in battle, they would collapse and decompose not long afterwards.

As I hinted in the chapters on Orcs, the absence of any discussion on the life-history of Orcs, just like the absence of anything very specific about their reproduction, tends to suggest either that Orcs were manufactured or had some kind of clonal reproduction in which the usual rules of life and death do not apply. What do I mean by this? The experience of a 'history' of life, incorporating a finite moment of birth, a process of aging and eventual death, implies the existence of individuals in a species that have a definite and identifiable genetic heritage. In clonal organisms, in contrast, this individuality is much less marked, because parents and offspring – indeed, entire populations – share the same genes. Clonal organisms can be seen as cells or components of one, multipartite, fissile, and effectively immortal superorganism.

The case of the Ents illustrates the extreme variation in life histories of Tolkien's 'speaking peoples.' Ents, like the trees they resemble, have a somewhat undefined life expectancy. They are capable of living to a very great age, although they tend to become somnolent and 'treeish,' and can suffer from the usual ills that wood is heir to, such as fire and axes. Treebeard himself is addressed by Gandalf as 'Eldest,' reported as the oldest living sentient being in Middle-earth, and having memories of walking in Arvernien and Nimbrethil, regions of Beleriand that perished at the end of the First Age (*Rings* III,4). This makes Treebeard 6,400 years old at the very least, and probably much older. As I shall discuss below in the context of Elves, large, long-lived creatures such as Ents tend to reproduce infrequently, a habit that exposes them to a large degree of extinction risk. Indeed, the immediate cause of the impending extinction of the Ents is their failure to reproduce, the Entwives having disappeared.

The Dwarves are described as a unique and separately sub-created race, different from Men and Hobbits, and their life-histories and habits are described in an intriguing note in *HOME* XII. This is where we learn about the large numerical excess of male over female dwarves, and that female Dwarves wore beards, making it difficult to tell the difference between males and females from a cursory inspection.

In terms of what we understand about evolutionary theory, these statements are contradictory. Species in which males and females look similar are also those in which the ratio of the sexes is likely to be even, not skewed as found in dwarves. It is in those cases of marked sexual dimorphism — when males and females look different — that the sex ratio deviates markedly from this. Humans are sexually dimorphic (the beardlessness of females is one aspect of this), but the sex ratio is very close to being even, most of the time; that Dwarves have such numerical inequality and show much less sexual dimorphism than humans is peculiar.

As regards life expectancy, we learn from the same source that Dwarves were much longer-lived than Men or Hobbits and could regularly live well into their third century, in which the only danger they were likely to face would be extreme corpulence (note the description of the aged Bombur in *Rings* II,1). Dwarves are, however, mortal, and, like Men and Hobbits — but possibly unlike Ents — their life expectancies tend to be fairly predictable. With the possible exception of the tradition of the reincarnated Durin, no single Dwarf lives for thousands of years.

Hobbits are very like Men. Indeed, Tolkien repeatedly makes it clear that Hobbits are really a variety of *Homo sapiens* (this is evident from the Prologue to *The Lord of the Rings*, as well as in several referencese to Men and Hobbits in *HOME* XII). Men and Hobbits share many of the same likes and dislikes, and there is much in common in their patterns of behavior and their family and social structures. Hobbits, however, live, on average, slightly longer than full-sized Men, with life expectancies of beyond a century, although Bilbo's age of more than 130 when he departed from the Grey Havens was exceptional.

Most Men in Middle-earth have the potential to live for the biblical span, if they do not die first from disease, starvation, or violence. The Dunedain — Men of Numenorean race – were once as short-lived as any other human beings, but were granted long life at the end of the First Age as a reward for fighting with the Elves in the first wars against Morgoth. In *Athrabeth Finrod Ah Andreth* (*HOME* X), Andreth, a wise-woman of the Edain, tells the Elf-lord Finrod of a tradition in human beings that they were once immortal, but had become corrupted by Morgoth and had suffered mortality as a result — a close parallel with the story of the Fall of Man in the Bible. The same source veers close to the explicitly Christian in its discussion of hopes that the One would come to Earth as some kind of redemptive process.

By the end of the Third Age the life spans of the Dunedain are very much diminished in comparison with their ancient glories, although they are still very great by ordinary human standards. Aragorn, for example, has the appearance of a rugged 40-year-old, so that it is easy to forget how old he really is. On March 1, 3019 — the day that he meets the reincarnated Gandalf in Fangorn — Aragorn celebrates his 88th birthday: he is, in fact, older than the aged King Théoden of Rohan.[1]

The Dunedain also age differently from ordinary Men. Rather than aging gradually, senescence comes upon them rather suddenly: as described in *Akallabêth* (in *The Silmarillion*), the Dunedain enjoy a long and hale middle-age after which they senesce very quickly. For this reason, they are granted the ability to expire quietly before they fall witless from their chairs. *The Tale of Aragorn and Arwen* (*Rings* A) shows the aging Aragorn (aged 210) accepting this same fate.

Given that evolution in Middle-earth tends to be Lamarckian rather than Darwinian, so that the benefits accrued by the Edain in ages past can be passed to their descendants, the variation in lifespan in the Men of Middle-earth is genetic – this is amplified by the state of things among Hobbits when considered as a distinct variety of *Homo sapiens*, among whose distinctive traits is a tendency towards long life.

Genetic variation in lifespan is to some extent true of Men in the real world, but in no case to we see the natural lifespan of *Homo sapiens* vary as greatly as it does in Middle-earth. Every person has the potential to live to between 70 and 100 years, although life expectancy at birth is strongly influenced by circumstances. People in war-torn parts of Africa have life expectancies at birth of fewer than 50 years; in the affluent North it is three decades longer. The increased prosperity of the United States has been matched by an increase in life expectancy since 1900 from 47 to 76 years, an increase of 62%. Nevertheless, it remains the case that only a very few humans can be expected to live beyond 120, no matter how blessed their lives — and there is no known instance in which human beings regularly live to, say, 200 or 300, as did the Numenoreans at the height of their power.

The limited variation of lifespan in modern *Homo sapiens* may, however, be a consequence of the fact that *Homo sapiens*, in its genetics, is notably homogeneous. There is more genetic variation in a single population of wild chimpanzees than can be found in the entire human race.[2] Homing in on humanity itself, there is vastly more genetic variation among people indigenous to the

t of Africa than among people in the entire rest of the
mbined — an artifact of the migration of modern *Homo
sapiens* from Africa between 200,000 and 100,000 years ago and the
subsequent rapid spread of the species over the face of the planet,[3]
extinguishing indigenous forms of humanity such as Neanderthal
Man and perhaps archaic relics of *Homo erectus*, as I discussed in
the preceding chapter.

The bottom line is that the genetics of modern humans is
surprisingly uniform, and this uniformity might also be characteristic of other facets of human biology, including lifespan. It could
be that the range of lifespan among humanity when considered in
its broadest sense might have been greater than it is now, more like
that of Middle-earth than the modern world in which the dominion of *Homo sapiens* is complete.

Evidence is beginning to accumulate for just these kinds of
differences in life history between *Homo sapiens*, Neanderthals,
Homo erectus, and other extinct forms of humanity. The evidence
comes largely from fossilized teeth, to a lesser extent from bones,
and from what we know of the evolution of childhood, perhaps
the single most important defining characteristic of modern humans.

Most mammals have a simple life history. They are born, they
grow rapidly to maturity, they reproduce, and then they quickly
senesce. Animals that have passed their reproductive peak are a
drain on resources and, therefore, die. This stark history is in tune
with the merciless logic of natural selection, in which animals
must strive to reproduce as quickly and as efficiently as they can,
and once they have done this, to leave the stage.

Humans, in contrast, are the odd ones out, bucking the evolutionary trend. In humans there is an extended period between
infancy and adolescence that we call 'childhood,' which lasts for a
decade or so. Children consume large amounts of resources in this
period, without making any immediate contribution in terms of
reproductive output. At the same time, human females, almost
uniquely, may enjoy a life expectancy of several decades after the
menopause, long after they have ceased reproduction. These facts
fly in the face of natural selection in that they impose a huge cost
on the species at large, in that there are many individuals that
consume resources without turning them into more humans —

humans may spend half their lives unable to reproduce. Logic suggests that humans ought not to exist, having been outcompeted by creatures with shorter lives and greater reproductive potential.

But there's more to it than that. The costs of childhood and old age are amply repaid by social benefits so powerful that they can be said to have made humanity what it is today – a species that, despite all expectations, has seen off all the competition and now dominates the Earth like no other single species in its 4.5 billion-year history.

How did we get to be like that? Various factors are involved, but they have become so intertwined that it is impossible to see which came first. One is the developmental state of children. Compared with the newborn babies of apes such as chimpanzees, newborn humans emerge in a relatively undeveloped state, re-quiring constant physical nurture. Of particular importance is the brain, which grows significantly in the first year of life. The brain is allowed to expand because the skulls of newborns are not fully knit together, and do not become fully rigid until a child is five years old or so. This brain expansion allows children to learn facts about the world, augmenting any inbuilt reflexes or instincts. The brains of most other mammals, in contrast, are fully formed relatively early in life, limiting the capacity to learn.

On the other hand, most mammals acquire all they need to know about the world relatively early in life, so their demands on their parents are fewer. The demands of human babies and chil-dren are much more intense, both because of their physical inca-pability as newborns, and the time required to learn and grow. In this situation, natural selection has favored the evolution of social collectives, typically extended families, whose members can share the burdens of raising children and in which children can learn from their non-reproductive elders.

Homo erectus evolved just over two million years ago in Africa, and was perhaps the earliest known species that we might regard as 'human.' This creature stood fully upright in a way almost indistinguishable from modern humans, although it had a much smaller brain. It was the first human species to move out of Africa, spreading across Eurasia from Spain to Indonesia and China. *Homo erectus* used fire and invented the hand-axe, that most emblematic icon of stone-age technology. Hand-axes are very distinctive tools associated with *Homo erectus,* and have been found all over the Old World.

But was *Homo erectus* really human? The answer, according to paleontologists Alan Walker and Pat Shipman in their book *The Wisdom of Bones,* is 'no.' Analysis of a near-complete juvenile skeleton of an early form of *Homo erectus* suggests that it grew much more rapidly than *Homo sapiens,* achieving maturity much more quickly, with childhood only a minimal interlude. *Homo erectus* may have looked human, but was possibly no more like us than any other social carnivore such as, say, a lion or hyena.

One might argue that technology — fire and stone tools — is evidence for a rational mind. However, the sameness of hand-axes, essentially unchanged for a million years, suggests that their construction was the product of instinct rather than any kind of technological application with which we are familiar. Modern human technology is always changing as new circumstances arrive. In contrast, *Homo erectus* made hand-axes in the same stereotypical way that birds make nests, or termites build mounds.

Corroboration of the nonhuman status of *Homo erectus* comes from studies on fossilized teeth. Because they are coated with enamel, the hardest substance produced by living organisms (see the chapter on *mithril* for more details on the concept of hardness), teeth are preserved in the fossil record far more commonly than other parts of the skeleton. What we know of fossils, including fossil humans, comes mainly from teeth. But today's teeth, of children and adults alike, are also subject to close scrutiny by thousands of dentists, all over the world, every day of the week. As a result, we have a detailed and extensive knowledge of the microstructure of tooth enamel, and how enamel 'growth lines' — laid down regularly, like the rings of a tree — track the development of individuals from childhood to adulthood. In particular, there is a close association between tooth development and brain development, allowing scientists to get a good record of the growth rates of the brains of extinct species, just by looking at fossilized teeth.

Briefly, teeth that grow quickly have bigger spaces between growth lines, implying a shorter period of growth, less time for a brain to each full size, implying a shorter childhood. Teeth with densely spaced growth lines, in contrast, imply slower maturation, longer childhood, and possibly longer post-reproductive lifespan. Teeth provide a solid, material gateway to understanding childhood, an otherwise evanescent character, yet vital to an understanding of the evolution of humanity.

Some recent research has shown that the trajectory of tooth development in *Homo erectus* was more like that in modern apes and extinct ape-men such as *Paranthropus* than in modern humans, suggesting that *Homo erectus* matured rapidly without the period of childhood associated with humanity.[4] From this one might speculate that *Homo erectus* had a life expectancy similar to that of extant apes: that is about 50 years.

Looking further afield, from humanity to other warm-blooded species (in particular, mammals), we can see that lifespan is connected, in a very general way, to attributes such as physical size, metabolism, and modes of reproduction. Small mammals burn energy much more rapidly than large ones, live for much shorter times, and tend to have large numbers of offspring that develop rapidly. Large mammals, such as humans, apes, elephants, and whales, have slower metabolisms and live for long periods. They have relatively few offspring, and devote a great deal of time and resources to their upbringing. Although this pattern sometimes disappears when examined too closely (for example, bats have similar sizes and metabolic rates to mice, but live ten times as long), and is modulated by other factors such as diet, we might speculate that the spark of humanity could not have emerged in a lineage of small mammals, such as rodents.

All these things explain why different species have different life spans. They do not explain why all of us, irrespective of whether we live to the age of two months or two millennia, eventually grow old and die.

Strange as it may seem, the important factor here is not death, but the process of aging or 'senescence' that leads up to it. In some ways, it is more useful to see death as the result of senescence, not an end in itself. Consider: the aged almost always die after suffering a long period of illness. They do not generally end their lives in Numenorean fashion, by collapsing without cause after a prolonged period of uniformly good health. Ageing takes a long time to happen. Even quite young people find their hair growing gray, their vision and hearing not quite as acute, their constitutions not quite as robust to over-indulgences as they once were. Older people suffer from progressive degrees of what can only describe as 'wear and tear' as their organs become less able to maintain themselves until, one day, the threshold of maintenance falls below some critical level necessary to sustain life. But why should this happen at all? Why don't we stay young forever, like an Elf? Clearly, it is senescence, not death, which we must explain.

As might be imagined, the causes of senescence are the subject of much research activity. A number of interrelated theories have been proposed to explain it, some of which are backed up by experimental research on various kinds of laboratory animal.

One theory seeks to explain the slow decay of bodily function in terms of genetics. With each year that passes, you outlive more and more of your peers who succumb to disease or accidents until, one day, you might wake up to find that all your friends and relatives have died. This is more than just fortune — you are clearly blessed with a constitution that is resistant to all the things that have killed everyone else. You have genes that resist disease.

There is a cost, however, in the accumulation of mutations that become apparent only after everyone else has succumbed. What do I mean by this? This point is best illustrated by example and experience. The average American born in 1900 would not have expected to have lived beyond 1950. In those days, infectious diseases claimed many more lives than they do now. Because we have largely subdued infectious disease and know a great deal more about nutrition and health, the same American born today might reasonably expect to see his 80th birthday. However, he might also expect to confront a range of problems that people in his grandfather's generation did not encounter, because they died from other things first. Cancer, Alzheimer's disease, Parkinson's, and other complaints of aging have waxed, as diphtheria, tuberculosis, and polio have waned.

Diseases of aging have a genetic component. Cancer, in particular, is believed to be triggered after a critical accumulation of genetic mutations, and with every year that you live, the chance of this trigger being activated increases. But why should this happen at all? Why do we not have the constitutions to survive these threats, too? Well, we might, but evolution does only what is necessary, not what is desirable – in the terms of raw nature, all creatures must do is avoid the hazards of life long enough to reproduce. For example, ninety percent of mice die in their first year, so there is terrific pressure on them to breed early. As far as the gene pool as a whole is concerned, there is not much to be gained by equipping the small minority of second-year mice with the genetic machinery required to ensure a life any longer than this. In evolutionary terms, elderly mice are expendable and can be allowed to fall to bits. The same is true, alas, for elderly people.

Another theory looks at the same problem from a different angle. It could be that those mutations that predispose elderly people to disease actually enhance their reproductive capacities when they are younger. In other words, there is a kind of genetic trade-off in which people who have children early and prolifically tend to have shorter lives. This seems to be borne out by demographic evidence. Women living in risky or unstable environments tend to have children sooner than women in more comfortable, stable surroundings: just like mice, the risk of early mortality is the spur to breed sooner rather than later. Conversely, there is evidence that women who choose to start their families later tend to age more slowly and live longer: it is no accident that these women tend to be wealthier and more educated than women who have children early. Looking at the wider picture, animals that delay reproduction and have fewer offspring tend to be long-lived, reflecting relative freedom from predation and risk.

These observations are, however, demographic, and say little about the underlying biochemical causes of aging. Many scientists agree that aging is connected with the products of metabolism. It seems to be a fact that the very biochemical activities inside your body that keep you alive will, ultimately, kill you. The reason is that many biochemical reactions produce toxic by-products called 'reactive oxygen species.' It may come as a surprise that oxygen, which we breathe to survive, is a highly dangerous and reactive substance. In most cases, however, oxygen atoms neutralize one another by getting together in pairs to create relatively benign oxygen molecules (O_2). However, the products of some biochemical reactions can produce combinations of oxygen atoms that are corrosive, damaging, and even mutagenic.

One way to minimize risk from reactive oxygen species is to reduce metabolism, and this may be one reason why long-lived animals tend to have slower metabolisms than short-lived ones. But in any animal, metabolism is dampened by eating less food, so reducing the amount of matter that the body is obliged to break down. This so-called 'caloric restriction' has been shown to prolong life in many species of laboratory animal. It is possible that the simple, subsistence diets enjoyed by some populations is connected with longevity.

Caloric restriction aside, the body has a number of mechanisms to combat reactive oxygen species, and repair the damage they cause. The laboratory roundworm *Caenorhabditis elegans* lives 44% longer than normal when treated with substances that neutralize reactive oxygen species.[5] As if to emphasize the adage that whatever doesn't kill you makes you stronger, a certain, constant exposure to reactive oxygen species — so-called 'oxidative stress' — primes these mechanisms and keeps them supple and responsive. In this way, aging can be seen as a diminution in our ability to rise to the challenge that oxidative stress imposes, and there is some actual evidence for this, in mice, if not in humans.[6]

It seems that aging is the result of exposure to chemicals, produced by our own bodies or otherwise present in the environment, that progressively degrade our tissues and might even cause cancer (caused, at root, by mutations in DNA that may be caused by oxidizing agents). However, it is highly likely that our ability to respond to oxidative stress is influenced by our own genes. This may explain why some people live longer than others no matter what, and why longevity runs in families.

What, you may ask, has this to do with Middle-earth? It is not for nothing that I have so far neglected the Elves, the subjects of this chapter's title, for my final task in this chapter is to address the scientific basis of Elvish immortality. If it were true that Elves really were immortal, then there would be nothing to discuss. However, it is not true to say that Elves live forever. Nothing in Tolkien is ever that simple: as Tolkien makes clear in unpublished essays such as *Athrabeth Finrod Ah Andreth* and *Laws and Customs among the Eldar* (both found in *HOME* X), Elves are not really immortal. Bound to the Earth, they live for as long as the Earth does, but this interval is implied to be finite. Elves live for a very long time indeed, but not forever. When Arwen cleaves to Aragorn and chooses a mortal life, she is not sacrificing immortality for mortality, but choosing one form of mortal life-history strategy over another.

That Elves are extremely long-lived rather than immortal ties in with everything else Tolkien tells us about their character. Like all long-lived creatures, Elves reproduce rarely (Elf children are not absent, but appear far less often than Hobbit or human children) and invest a great deal of care in their offspring. More interestingly, Elves do not suffer from disease or aging, although may die of grief or by accident. This privileged existence will itself predispose Elves to live long lives, given what we know of the trade-offs that exist between early reproduction, health, general risk, and longevity.

This suggests that Elves have a genetic constitution that is not only resistant to infectious disease, but also those diseases of aging, such as cancer, that seem to be related to a diminution of the body's resistance to oxidative stress. Elvish versions of DNA-repair enzymes, as well as catalase or superoxide dismutase those enzymes whose task it is to combat reactive oxygen species — must be potent indeed. Elvish lives are also prolonged by caloric restriction alongside slow metabolism. Tolkien makes an issue of the capacity of Hobbits, Dwarves and Men to indulge themselves in the pleasures of the table, but this is much less true of Elves, who, while they enjoy the occasional banquet, largely seem to exist on a diet of water, *lembas,* and thin air (*Rings* III,2) and can go for long periods without visible means of sustenance or even sleep. This suggests a highly efficient metabolism in which the production of reactive oxygen species is kept to a minimum. Because of this, Elves always manage to look young and beautiful, signs of age being revealed only in their eyes. Aragorn may have turned 88 in the War of the Ring, but Arwen[7] is 2,778, and her grandmother Galadriel is 8,440.[8]

The fading or death from grief that is one of the two possible causes of Elvish mortality is, in this model, easy to explain – it results from a dramatic lowering of resistance to oxidative stress in which the body is quickly consumed, as if on a pyre: *The Silmarillion* (chapter 10) records how quickly the body of Fëanor, perhaps the most vital of all the Elves, burned away once his fiery spirit had fled.

16. GIANT SPIDERS AND 'MAMMOTH' OLIPHAUNTS

No Tolkien fantasy is complete without a gigantic spider or two. If Bilbo fights off the giant spiders of Mirkwood in *The Hobbit*, Frodo and Sam have a harder time of it with the altogether more sinister Shelob in *The Lord of the Rings*, and the arachnid monster Ungoliant, the progenitor of them all, crops up repeatedly in *The Silmarillion*. Tolkien's preoccupation with giant spiders is never satisfactorily addressed. The canonical story as related in the *Biography*, that Tolkien was stung by a large spider while a toddler in South Africa, might not, in my opinion, be sufficient to explain the prevalence of huge spiders in fiction written decades later.

I'd like to suggest that giant spiders provide remarkably good monsters in a particularly Tolkienian mode. On the most basic level, spiders are creepy, playing on the common horror of arthropods (jointed-legged animals) in general. Our horror at the monster in Ridley Scott's *Alien* evokes precisely this distaste. Tolkien would have been familiar with the giant ants and so on that were once the staples of pulp science fiction — spiders, unlike trolls, dragons or even balrogs, have an inhuman quality that makes them especially pitiless adversaries, especially if — like Wells's disgustingly squishy Martians in *The War of the Worlds*, they are also gifted with intellects 'vast and cool and unsympathetic.'

More deeply, Tolkien's monstrous spiders, their thirst for light forever unquenched, and their production of webs of impenetrable gloom, cast them as agents of a disturbingly active kind of darkness, exemplifying what Shippey (in *J. R. R. Tolkien: Author of the Century*) has identified as a fundamental ambiguity in Christian theology. That is, the nature of evil: is it an active agent in its own right, or simply the absence of good? Is the gloom emanating from Torech Ungol just the absence of light, or something more potent, an 'unlight,' an emanation that clogs the senses and inspires terror? In this respect, the pointed neutrality of both Ungoliant and Shelob — they are associated with, respectively, Morgoth and Sauron, but, uniquely in Tolkien's bestiary of evil, retain their independence — may be significant in this respect.

Of perhaps more concern here is whether it is possible to have spiders that are as large as Shelob, and in this chapter I look at this question as well as another notable instance of Tolkienian gigantism – that of the mighty Oliphaunts, the Mûmakil of Harad. I close by asking whether there might be some literary reason for the invocation of gigantism in Middle-earth, more than just for the sake of awe and spectacle.

What determines how big animals can grow? This question is far simpler to ask than to answer. Indeed, it is one of the most active and controversial topics in science today.[1] We do know, however, that size is connected with metabolism — that is, the way that an animal uses energy — and the degree of interaction between an animal and the environment in which it lives. This makes a good place to start, because it turns out that size can be limited by the need for an animal's insides to interact with the world in which it lives.

As an animal gets larger, its volume expands at a greater rate than its surface area. The reason is that area scales with the square of length, whereas volume scales with the cube, so that the more an animal grows, its volume will get proportionately larger in relation to its area. This constraint, in fact, is a simple consequence of geometry, and applies to any object, provided that it maintains the same shape as it grows.

To see this effect in action, think about a child's set of building blocks. Imagine that all the blocks are cubes of the same size, and that each cube measures one inch along each edge. It's easy to see that a single block has a surface area of six square inches, and a volume of just one cubic inch. This gives a ratio of surface-area-to-volume of $6/1 = 6$.

Now, imagine stacking the cubes into a larger cube with a length on each edge of two inches. To do this, you'll need to make a pile of eight cubes, creating a larger cube with a total surface area of 24 square inches and a volume of eight cubic inches, giving a surface-area-to-volume ratio of 24/8 =3, that is, half that of the smaller cube.

The larger cube has proportionately more *insides* than the smaller cube. If this all sounds a little confusing, try this analogy: you can get many more passengers inside a Boeing 747 than in a smaller 737 — but proportionately more 737 passengers get window seats.

This has dramatic consequences for animal life. Animals need to absorb, consume, or otherwise ingest substances vital for life, such as air, water, and nutrients, and they must also void wastes. Very small creatures are able to exchange water, nutrients, and waste directly through their surfaces, simply because they are small, and their surface areas are large in relation to their volumes.

We human beings are far too large to be able to exchange all the substances we need through our skins, but evolution has found ways around this by increasing the surface areas of internal structures, using our greater volume as a way of accommodating organs with enormous surface areas, rolled up tight. Our digestive systems allow us to ingest concentrated dollops of nutrients and digest them in a very convoluted structure, the small intestine, which, if completely rolled out, would have the surface area of a tennis court — vastly greater than our skin. Our lungs are not plain sacs, but massively complicated branching networks of tubes and sacs that together have an area in the tennis-court range.

Most of our internal organs are constructed so that they have large surface areas crammed into small volumes, as a way of subverting the geometric tyranny of the surface-area-to-volume ratio. The subversion is, however, as geometric as the constraint, and relies on the let-out clause that volume increases relative to surface area only if the overall shape is maintained. It need not apply if shape changes to compensate. To return to the example of the child's building blocks, if you take the cube made out of eight blocks and rearrange the constituent cubes into a tower eight blocks high, it will still have a volume of eight cubic inches, but its surface area will have grown from 24 to 34 square inches, giving a surface-area-to-volume ratio of 34/8 or 4.25 – much greater than the same volume arranged as a cube.

Heat, however, is a much more difficult commodity to exchange, because it cannot simply be bundled up into packets, like nutrients, and to a lesser extent air and water. In fact, the more complex a system becomes (by, for example, bundling a lot of surface area up inside and busily shuffling substances across it), the more problematic heat becomes, because moving substances across surfaces uses energy, which generates heat.

As far as heat is concerned, the limiting factor is the skin itself, the interface between the animal and the outside world. All animals generate heat to some extent, and they also lose heat: the so-called 'cold-blooded' animals, which do not have a set running temperature, gather and lose heat by basking in sunshine, or being active at warmer or cooler times of day. Warm-blooded creatures such as mammals and birds have set running temperatures that are typically higher than that of the ambient environment. If the animals are small, they will tend to lose more heat than they gain, and so will have an insulating coat of fur or feathers. If they are large, however, the bigger problem is heat loss, because as animals grow it becomes harder for the heat generated in their voluminous bodies to reach the surface. This is one reason why large animals such as elephants tend not to be furry, and their capacity to shed heat is enhanced by having large blood-filled ears, which act as radiators, as well as heat-losing habits such as swimming or wallowing in mud.

Metabolism, however, comes to the rescue. As I discussed earlier, larger animals tend to have slower-running metabolisms than smaller ones, and this may allow animals to grow larger than they would were this not the case. The relationship between size (or, more accurately, mass) and metabolism is a matter of much current debate.

Given the relationship between size, area and volume just discussed, you'd expect that the relationship between mass and metabolism would depend on the *square* and *cube* of the mass. Mathematicians might express this relationship as

$$R \propto M^2/M^3$$

where R is the metabolic rate, M is the mass, and the symbol '$\propto$' means 'is proportional to.' In words, the expression 'M^2/M^3' means the square of the mass divided by the cube of the mass, which is another way of putting the surface-area-to-volume ratio. This can be tidied up to give the following expression

$$R \propto M^{2/3}$$

Which, when translated, means that metabolic rate is proportional to the value of mass raised to the power of two-thirds. It may seem strange to see powers expressed as fractions, but this is simply an arithmetic consequence of dividing the square of a value by its cube. It also expresses, very simply, the concept that as animals get bigger, their metabolisms do not accelerate to match their sizes. In other words, larger animals have disproportionately slower metabolisms.

The problem is that this view now seems to be overly simplistic, a consequence of thinking about living things as if they were solid masses, like stacks of children's' building blocks. A great deal of experimental evidence has now accumulated to show that the scaling relationship is rather different, closer to

$$R \propto M^{3/4}$$

At first sight this seems extremely strange – where does this 'fourth' power come from, this extra dimension of size? The reason could be a simple consequence that many of the volumes inside our bodies are used to accommodate greater expanses of surface area than you'd expect simply from looking at the animal from the outside. That is, our insides have a kind of 'extra' dimension, rolled up tightly: we are very much more than masses of homogeneous building blocks.

Nevertheless, whatever the precise relationship between metabolism and mass, it is clear that larger animals have disproportionately slower metabolisms than smaller ones, and, as I discussed before, this is connected with greater longevity, fewer offspring – and also greater rarity. The resources available in any given environment are finite, so there will be a connection between an environment and the sizes and abundances of animals it can support. Generally, there will be proportionately fewer larger animals in a given environment than smaller ones.

This phenomenon is particularly acute when large animals become marooned on islands as a consequence of having been cut off from the mainland by rising sea levels: a sudden decrease in land area imposes a great value on small size, because smaller animals live more economically than larger ones. The fossil record is spotted with examples of dwarf, island versions of creatures that once lived on the mainland. Dwarf deer once lived on the Channel Islands (in the English channel); a pygmy elephant once roamed the island of Malta, and a pony-sized mammoth once inhabited Wrangel Island, once a promontory of the Ice-age subcontinent of Beringia, which became isolated when sea levels rose at the end of the last Ice Age.[2]

One limiting factor on size, therefore, may be that environments cannot support animals greater than a certain size, together with the possibility that very large animals breed sufficiently rarely for their populations — already small in number — to be at constant risk of extinction.

There are, however, other constraints. One has already been mentioned, and that is heat. The largest land animals ever to have walked the Earth were dinosaurs, some of which might have had masses in excess of 80 tons. The question of whether dinosaurs were warm- or cold-blooded is largely unresolved. However, copious generation of heat from metabolic processes might have been problematic for the largest dinosaurs whatever their metabolism, and this would have been exacerbated had they been warm-blooded. We know for certain, however, that mammals are warm-blooded, and no mammal has ever achieved the size or mass of a dinosaur.

The largest mammal alive today is the African elephant *Loxodonta africana*, the largest examples of which may weigh 10 tons. This was exceeded in the relatively recent past by the mammoth *Mammuthus trogontherii*, the largest known species of elephant, which may have weighed as much as 16 tons. This species was a precursor of the smaller woolly mammoth *Mammuthus primigenius*. Larger still was the extinct hornless rhino *Indricotherium transouralicum* (also known as *Baluchitherium*), estimated to have weighed between 15-20 tons. Larger estimates may have been in error, having been based on inadequate fossil material.[3] In any case, *Indricotherium* may have been much more rangy and spindly-looking than an elephant – think of a giraffe threaded through a rhino. Presumably, such large creatures would have had ways of losing excess heat. Another limit to size in mammals might have been reproduction. It could be that carrying very large embryos imposes a great burden on very large mammals that would not have been a consideration for dinosaurs, which laid eggs.

Tolkien described the Oliphaunts encountered by Frodo and Sam in the woods of Ithilien (*Rings* IV,4) as gigantic elephants, larger than those found today, but they were possibly no greater than, say, *Mammuthus trogontherii*. The Oliphaunts in Peter Jackson's films of *The Two Towers* and *The Return of the King* are very much larger than this, comparable with large dinosaurs, and these would be biologically problematic. In brief, it would be very difficult to scale an elephant to the size of a dinosaur and have it look and behave just like an elephant. Elephants the size of dinosaurs would have looked more like — well, dinosaurs, or at the very least not as much like elephants as they are shown in the movies. The ancestors of elephants were four-square quadrupeds, and this imposes certain structural constraints when animals become very large. The legs of dinosaur-sized elephants would have been much thicker, more column-like, and altogether less bendable than those of regular elephants. They would not have moved very quickly or reared up on their hind legs. Dinosaurs, on the other hand, achieved their large size in part because their ancestors were lightly built bipeds, with a bone structure more like that found in birds, and capable of very rapid growth. This lightness may have allowed dinosaurs to grow far larger, while remaining more active, than might have been possible for mammals.[4]

And although the metabolism of gigantic elephants would have been slower than that of regular elephants, they would still have been warm-blooded, making heat loss a problem – their ears might have been even larger in proportion to their heads than is the case for African elephants. When thinking of an appropriate model to scale up, think Dumbo rather than Jumbo. So, whereas Tolkien's Oliphaunts were just about plausible, Jackson's may not be.

Tolkien runs into trouble with his giant spiders, however. The spiders in *The Hobbit* are comparable in size with Bilbo and the dwarves, whereas Shelob was even bigger, able to crush Sam or Frodo entirely beneath her body. The largest spider alive today is the Goliath bird-eating spider (*Theraphosa leblondi*), a species of tarantula from South America. With a leg span of about 12 inches, it can cover a dinner plate at full stretch, but the body of this nevertheless formidable and rather scary animal would not cover a dollar bill – hardly a match for Shelob.

Spiders and other jointed-legged creatures were, however, bigger in the past. In the Carboniferous period, around 300 million years ago, the world's forests boasted dragonflies the size of crows; a two-meter millipede called *Arthropleura*, meter-long scorpions and a spider, *Megarachne*, which had a body the size of a bagel and a 20-inch leg span. Nobody knows why arthropods were larger in the past, and specifically in the Carboniferous Period, although a widely held view is that the atmosphere at that time contained appreciably more oxygen than today, allowing these creatures to grow larger.

Nevertheless, none of these arthropods grew large enough to match even a moderately sized mammal, still less a dinosaur. The reason could lie less in absolute size than in the way arthropods are made. Vertebrates, including dinosaurs, elephants and people, are supported by internal skeletons of cartilage and bone, on which a variable amount of flesh can be draped. Arthropods, on the other hand, are built inside-out — they wear their skeletons (made from a protein called chitin) on the outside, like suits of armor. The only way that arthropods can grow is to shed their skeletons, puff themselves up, and secrete a new skeleton: the animal remains extremely vulnerable while this happens, and unsupported, unprotected arthropods more than a certain size may not be able to support their own weight without damage. Giant spiders, then, are at least as impossible, biologically, as dinosaur-sized elephants.

We can be thankful, however, that Tolkien (and Peter Jackson, for that matter) ignored biology to give us such memorable monsters of page and screen. The question is raised, however, why so many of Tolkien's monsters were large and, in addition, larger versions of creatures seen today. Apart from giant spiders and elephants, there were (as discussed earlier) giant eagles, giant pterodactyl-like Nazgûl-birds and giant – well, giant giants, in the form of trolls and ents. The clue comes, I think, in a rather buried homage to Earth's prehistoric past. Tolkien was certainly aware of the wonders of the fossil record, as evidenced by his rather sidelong remarks about 'pterodactylics.' Given that Middle-earth was meant to represent our own world at some unspecified moment of prehistory, Tolkien perhaps thought it fitting to populate it with gigantic creatures with an air of mystery or antiquity about them. The Oliphaunt is a case in point: 'oliphant' is an archaic version of the word 'elephant' from days when, to westerners (and certainly Englishmen), elephants were associated with faraway places and were so exotic as to be mythical, and, if seen, their size would have made an even greater impression than they would today.

Another key point, to my mind, is that most of these gigantic inhabitants of Middle-earth are described either as relics from a forgotten past (Shelob), as being on the verge of extinction (the Ents), or both (the Nazgûl-birds.) This ties in very well with what we know of the vanished world of the relatively recent geological past. No more than thirty thousand years ago, mammoths roamed the northern world. The Americas were inhabited by giant ground sloths bigger than elephants, and armadillos the size of cars. Deer with gigantic antlers roamed the plains of Europe, and rhino-sized wombats lived in Australia. All of these wonderful creatures have vanished. The arrival of meat-hungry humanity might be connected with the recent disappearance of most animals larger than a Labrador retriever, although changes in climate might also be responsible.

Much of the research on the extinction of the so-called 'megafauna' is relatively recent, and Tolkien would not have been aware of it. Nevertheless, the fact that Tolkien has the extinction of much of the megafauna of Middle-earth coincide with the start of the Dominion of Men has a stark and modern resonance.

17. INDISTINGUISHABLE FROM MAGIC

There are two occasions in *The Lord of the Rings* in which the Hobbits ask the Elves for a demonstration of Elvish 'magic,' only for the Elves to profess puzzlement at the word (*Rings* II,7; II,8). To the Hobbits, the term 'magic' is a catch-all term for phenomena beyond their clearly limited experience of technology, extending from artifacts that are plainly the products of technology, such as Gandalf's fireworks and 'magical' Dwarf-made toys from Dale, all the way to objects and phenomena that would seem magical even to the twentieth-century reader, such as doors controlled by magic spells, swords that glow in the dark, *palantíri*, and the Ring.

What Tolkien is telling us is that magic is not something that is supernatural, beyond the dictates of nature: rather that the way we perceive magic depends very much on our own technological sophistication. Hobbits, ignorant of any technology much beyond the water-mill, would regard gunpowder as magical. Modern readers, however, know very well that such trickery is neither more nor less prosaic than regular high-school science: a Hobbit visitor to a modern chemistry class would, no doubt, see it as sorcerous as a potions lesson at Hogwarts. Considered in the same way, the kind of magic routinely performed by the Elves seems magical to us only because we do not understand its physical basis. To paraphrase the famous aphorism of the science fiction author Arthur C. Clarke, any sufficiently advanced technology would seem indistinguishable from magic.[1] The bottom line is that there is no such thing as magic in Middle-earth: what seems magical to us is the product of Elvish technology that is far more sophisticated and mature than our own. Elvish magic in latter-day fairy tales is technology through the eyes of uncomprehending witnesses, debased by generations of repetition.[2]

Our own technology has advanced to a degree that we can begin to sketch how some of the products of Elvish technology might have worked, at least metaphorically, and I have already attempted this in the chapters on the *palantíri*, the Silmarils, and *mithril*. In this chapter I discuss some more, as a way of illustrating the general culture of the Elves, in that their technology had become so perfect that it was essentially contiguous with nature, working alongside it rather than against it. I pursue this theme to its logical conclusion to show that mankind can attain harmony with nature not by reverting to a never-land of rustic simplicity, but by modifying the world to such a degree that the lines between humanity, nature, and technology become blurred.

To understand the ever-changing relationship between people and technology we must first ask what technology is *for*. The answer is very simple: technology exists so that we can control our environment far more efficiently than we could unaided, with our bare hands. The simple technology of clothes, fire, animal-hide shelters, and stone tools allowed our ancestors to venture out of the tropics and penetrate environments never inhabited by unclad apes. However, technology is hindered at every stage by the fact that the devices we use to control our surroundings — which act as interfaces between ourselves and the environment — are imperfect. A bear-hide suit might protect you against winter storms, but would be hopeless in space. But even space-suits leak.

As anyone who uses modern gadgets such as computers knows all too well, the technology we use to save us time is often a hindrance as much as a help. Before we can place a call with our cell phone, we must learn how to use it, register with a network, and press a lot of buttons, and even then we do not always get through. However, placing a call is much easier and cheaper than it was even a generation ago[3] — the irritation comes from less well-tried technologies that accompany cell phones (digital cameras, messaging systems, wireless networking, and so on) that we, as customers, are effectively road-testing, but which people in five years time will use without a second thought.

When we think of the improvements in technology as regards power, affordability and ease of use, what we are *really* thinking is how technological advance can be measured by the degree to which it can facilitate the control we can exercise over the environment, by making itself so simple to use that it is, in effect, invisible. As I write, my pocket is cluttered with a personal digital assistant (PDA) and a cell phone, which can, if I spend some time poring over menus, communicate with each other, with my wife's PDA, and our home computer. My up-to-the-minute phone and PDA share pocket space with my house keys (representing the very latest in medieval *smipes orpancum*) and loose change (which has been with us since remote antiquity.)

But how much easier it would be were I to have a device no larger than a grain of rice — perhaps worn as jewelry, or even implanted — that would store everything I need to know, open every door, effect every transaction, and could be accessible by some kind of technological telepathy. I could contact people and access information as quick as thinking, with no buttons or menus or screens in sight, but just as if the technology was a part of me, an extension of my own senses.

This is not as science-fictional as it sounds. A device called the cochlear implant is a kind of interface that restores hearing to the deaf, and work is already progressing on ways to allow people, in particular the profoundly disabled, to interact with computers by way of neurological implants that take signals directly from the brain.[4] It is not beyond the realms of possibility that such technology would, one day, reach the mass market, at which point we would potentially become as one with our technological environment, and the tedium of getting our gadgets to do what we want them to do would, hopefully, evaporate.

Now, let us imagine what our lives might be like were we all to have such technology. Human beings would become independent as they become more interdependent. We could work wherever and whenever we wanted, spending more time in pursuit of leisure-time activities or, perhaps, just getting back to nature. Old-fashioned technology such as keys and money, not to mention phones, PDAs, and computers, would become redundant. No door would remain locked, international barriers would collapse, and possessions would be held in common, because everything of value would reside as decentralized information, accessible to anyone. Each one of us would be a node in the worldwide net.

Such idylls have been imagined before, of course[5] — we once thought that the advent of the internet and affordable computers would allow people to stay at home in droves, telecommuting while they spent less time in their cars or on trains, and more time with their families or playing golf. One of the reasons this hasn't happened is, simply, because the technology is unready: it has too often become an obstacle rather than an aid. The much-vaunted 'paperless office' hasn't happened because we do not trust computers sufficiently to archive important information (often with good reason.) None of this should be seen as a barrier to truly unobtrusive technology that becomes so much a part of us that we take it for granted.

The Elves at the end of the Third Age already have an equivalent level of technological sophistication, and have had it for millennia. Because of this technology, apparently wild woods can house sizeable populations of Elves who appear to have very little means of support, habitation or transport, and yet who have access to large amounts of information, some of it only available remotely. That they can communicate by a form of telepathic rapport is explicit (*Rings* VI,6). Of course, I do not mean that Elves jacked themselves into Middle-earth 'net' in any way that is directly equivalent to my own technological prognostications. That having been said, however, this kind of seamless technology explains a great deal about the Elves, their culture, and their behavior, as well as resolving the apparently 'magical' aspects of their technology.

An important aspect of truly sophisticated technology is that it is invisible. My favorite way to approach this problem is by way of an analogy with stereo systems, in which there seems to be a direct relationship between price and sophistication on the one hand, and obtrusiveness on the other. Cheap stereos are often brash, dominating a room with big speakers and flashing lights, as if they were statements of just how much you can get for your money. Expensive stereo systems, on the other hand, tend to be near-featureless boxes that can be hidden away, communicating by wireless technology with speakers that are either concealed or which look like something else — flower pots, perhaps, or oil paintings. On entering a room, you'd never know the stereo system was there, unless your host activated it with a spoken command, and beautiful music would instantly fill the room, but from no source that you could easily identify. It is a small step from this (or, perhaps, no step at all) to doors that remain resolutely shut until commanded to open with the appropriate magic spell. Open sesame: say 'friend' and enter. Theoretically, the cell phone in my pocket responds to my voice (if I could only work out how to program it), and speech recognition software is now commonplace. Spells to open or close things are no longer as magical as they would have been to Tolkien's audience of half a century ago.

That advanced technology should be unobtrusive says nothing in particular about how it is used by the Elves, who (as far as we know) do not have electricity, let alone powerful hi-fi systems. The Elves at the end of the Third Age have gone far beyond that. To be sure, the Elves had an 'industrial' phase that started with Fëanor in the First Age and culminated with the Elven-Smiths of Eregion in the Second, but partly as a result of the disastrous wars with Morgoth and Sauron consequent on such industrialization, Elvish technology became far more organic and less obvious. The major theme of Elvish technology is that it harnesses and exploits nature without it even noticing, rather than standing in opposition to it.

This is an important point, for in it is expressed many of the contrasts between the Elves, on one hand, and the forces of evil on the other. The darkness of Morgoth and Sauron is augmented by the unlight of the smokes of furnaces, and the pollution and waste that heavy industry leaves in its wake. According to Treebeard, Saruman is going that way too, his mind full of wheels and engines (*Rings* III,4). This is the kind of technology of which Tolkien clearly disapproved, and matters are only made worse when industry is used not just in opposition to nature, but to ape it. The generation of Orcs in 'mockery' of Elves is more than an insult: it is a self-acknowledging mark of technological inferiority, because everyone — including Morgoth — is aware that he is unable to use the brute force of industry to make a sentient being anywhere near as well-crafted as an Elf. Orcs, and the industry that spawned them, are cheap knock-offs. It is Elvish technology that is the real thing — technology that is all-pervasive but very hard to see, because it is hidden from us.

As I showed earlier, Tolkien disliked the kind of mechanical gobbledygook required to support a fantasy story, and so he is not likely to offer us any easy clues about how Elves perform their 'magic,' even though he admits, through their own voices, that what they are doing is not 'magical' but a normal part of Elvish life. We *know* that the Elves can communicate in thought, but we are, of course, offered no details; we *know* that Elves make beautiful and useful things with remarkable properties, but we are never explicitly shown how; we *know* that Elves live healthily and well, but we are never shown how this lifestyle is supported — in contrast with Orcs and the forces of darkness, there are no obvious farms, factories, or armies of slaves working the fields.[6] This absence does not mean that the technology is not there.

Before I go any further, a word of warning is apt: could I be attributing stratagems and devices to Tolkien that he never intended? Did he always mean Elves to be simple, rustic folks in a prelapsarian state of grace, albeit with genuinely supernatural powers? Not necessarily, for there are a couple of occasions, at least, where he almost gives the game away.

For one, there exists an unpublished note (in *HOME* XII) on the manufacture of Elvish 'waybread' or *lembas*. Although the note is mainly concerned with the purity of the strain of wheat used to make *lembas*, rather than advancing anything explicitly technological, it is notable that Tolkien mentions the production of stands of wheat in a society in which farming appears either absent or takes place offstage. In contrast to Hobbits, nowhere in *The Lord of the Rings* do we see the honest agrarian toil of Elves.

The second instance is more remarkable, and concerns Sam's want of a bit of rope. In *The Lord of the Rings* as published, Sam discusses ropemaking with the Elves, mentioning his family connection with the craft, and how much he would value a hank of cord. When the party leaves Lothlórien (*Rings* II,8), they find coils of Elven rope in their boats, which, they are told, is made of a substance known as *hithlain* whose nature is not made explicit, although the name in Sindarin might be translated roughly as 'threads [of] mist' – which, while descriptive, gives us no clue about its manufacture. In fact, the Elves say very plainly that they do not have the time to tell Sam how *hithlain* ropes are made.[7]

However, early sketches for the Lothlórien episode (*HOME* VII) have the Fellowship actually seeing Elves at work at broidery and carpentry, and watching how ropes are spun from a silvery fiber derived from the bark of the *mallorn* tree. To have included this in the published work would have destroyed the mysterious atmosphere of Lothlórien at a single stroke, which is presumably why Tolkien omitted it. But that he was thinking along these lines at all suggests that he had some conception of Elvish technology — and that he chose to hide it.

Once this conceit has been exposed, however, the scales fall from our eyes and it is plain that Elvish technology is all around us – in the trees and grass, the beasts and birds, and in the very air we breathe. As the Fellowship approaches Hollin (*Rings* II,3), a desolate region long-abandoned by the Elves, we are told that some Elvish goodness clings to the land nonetheless, and that the land does not quickly forget their influence. This is no mere 'magic,' for we are told that the Elves have none. The reason why landscapes remember the Elves is that the Elves have actively modified the landscape to their use, but in such a sensitive way — and using such subtle means – that we, the non-Elvish observers, have no reason to think that the landscape is anything but natural, albeit that the 'goodness' of the Elves somehow still clings to it.

I can think of two other examples in which people are shown to be deceived into thinking their surroundings are natural, when they are not. The first is fictional, and comes from *Contact*, the only novel by the late planetary scientist Carl Sagan. When the heroine, the astronomer Eleanor Arroway, finally gets to meet the alien, the creature tells her that the starry sky she has studied all her life is not the wilderness she had always taken for granted, but a landscape that had been heavily modified over hundreds of millions of years by cosmos-spanning civilizations, some of a power that awe even the alien himself.

The second example is real, and comes from the work of the anthropologist Clark Erickson of the University of Pennsylvania. Studying a network of raised paths that spread extensively throughout the Bolivian Amazon, Erickson realized that they were not random or 'natural,' but the man-made boundaries of pools formed during seasonal inundations, constructed in such a way as to funnel fish into basketwork fish-traps placed in strategic gaps in the raised paths. Pools of standing water would form adjacent to the fish traps, trapping sediment and allowing the growth of palms and other plants that could also be exploited by the dry-land fisherman.[8] These palm-fringed pools look entirely natural until and unless you are told otherwise. This and other work has made it clear that what we think of as the wilderness of Amazonia is, in fact, anything but — the jungle has been modified for thousands of years by its human inhabitants. We, accustomed to factories, or at least to neat fields full of wheat or cows, see only a jungle — the Amazonians see a market garden, modified for their use.

By analogy, the Elves are a hybrid between Sagan's celestial engineers and Erickson's landscape fisherman. We should never take the easeful woodland idyll of the Elves at face value, for over the course of millennia, Middle-earth has been modified by their presence in all kinds of ways that one might not expect. It could be that many of the things that look 'natural' in Middle-earth are actually the products of Elvish technology. For example, we hear from Treebeard (*Rings* III,4) that it was the Elves who woke the trees from slumber and taught them to 'speak.' In one sense, this is the story of Genesis taken further and inverted to reflect the goodness of the Elves, in contrast to that of Men — whereas Adam is told to name the flora and fauna in order to dominate them (Genesis 1, 28; 2, 19-20), the Elves encouraged the flora and fauna to give *themselves* names, and thus be set free.

However, it could be read as an echo of a legend in which the Elves refashioned trees into Ents, that is, the Ents are as artificial in their genesis as Orcs, but made in celebration rather than mockery. The sentience of many of Middle-earth's birds and beasts — the crows and thrushes of the Lonely Mountain in *The Hobbit* spring to mind - could be the result of Elvish intervention. One could go overboard here, especially as Tolkien himself speculated (*HOME* X) that the sentience of otherwise insensate beasts might be conferred by their being animated by spirits.

Nevertheless, the point is made, and few can doubt the dramatic and conscious influence of the Elvish inhabitants on the flora and fauna of particular regions, notably Lothlórien, characterized by the *mallorn* tree - a species found nowhere else in Middle-earth and which has a particular association with the Elves, as if the species, tree and Elf, were symbiotic in some way. While in Lórien,

Sam remarks that the association between the Elves and the Golden Wood is so close that it is impossible to decide if the Elves made the land, or *vice-versa* (*Rings* II,7).

There is a curious lesson here for environmental campaigners who see in Tolkien a kind of mentor. To be sure, Tolkien objected to the kinds of heavy industry and urbanization that destroys rural landscapes and wilderness. However, it seems clear that the Elves, because of their close ties to the Earth and their sensitivities to its needs, have modified the landscape more thoroughly than industry ever could, if less obviously — what we think of as good, in a landscape, even idyllic, is the result not of the wilderness itself, but of its enhancement conferred by the presence of Elves.

This kind of 'organic' technology is technology nonetheless, and those who are concerned with the effects of technology on the environment must ask themselves whether it is better to create a technology that is obvious and local, or a technology that is invisible but potentially far more powerful, because it is closely connected with nature — modifying what is already there, rather than imposing something from without. It would be the Elves, not Men, who would be the first to adopt the genetic modification (GM) of organisms — the difference between Elvish GM and the human variety is that the Elves would use it for the improvement of organisms for their own sakes and that of the Earth, or simply for the joy of being able to perform such acts, and not for any kind of profit or domination.

There is a distinction here, central to Tolkien's legendarium, but which is often misinterpreted. Technology itself is neutral, but can be made good or bad by usage. By the same token, technology – that is, the modification of our world to facilitate our control over it – need not be inherently a bad thing. We will become like Elves not by dominating it with smokestack industries, but in a more subtle, organic way that has the potential to be more far-reaching.

It is characteristic of the Elves that in seeking to solve a problem they should first find their inspiration from nature rather than their own invention. Raised in the darkness before the Sun and Moon, and yet enamored of starlight and illumination generally, the Elves soon sought ways to create artificial light. The Elvish lanterns seen throughout Tolkien's fiction are neither candles nor flaming brands. Nor, obviously, are they powered by gas bottles

y. I suggest that they are fuelled by bioluminescence – pensity of certain organisms to emit light, either through emical reactions or, in the cases of some micro-organisms, as a result of the physical deformation of their cell surfaces. Most bioluminescent organisms are found in the sea, and range from bacteria to deep-sea fishes. On land, bioluminescence is found in fireflies and in certain decay bacteria (the latter, presumably, being the source of the ghastly light emanating from Morgul Vale.) Very crude Elvish lanterns might employ some kind of bacterial bioluminescence, but I suspect that they would exploit preparations derived from bioluminescent organisms either on their own, or in some genetically engineered context.

One of the most useful and intriguing research tools to emerge from laboratories in recent years is the so-called 'Green Fluorescent Protein,' or GFP.[9] This was discovered as long ago as 1962 in a bioluminescent jellyfish *Aequorea*, as a companion protein to the luminescent jellyfish protein aequorin. Pure aequorin emits a blue light, but GFP efficiently turns it green. The jellyfish GFP gene, that is, containing the genetic instructions for GFP, was isolated and cloned (that is, copied into a different creature) in 1992, and it was subsequently shown that the gene produces a fluorescent protein even when transferred into an organism of a different species – that is, you could use GFP to make an organism glow in the dark, and no jellyfish need be involved.

GFP has proven amazingly useful to genetic engineers trying to assess the success of moving a gene from one organism to another. By linking the transferred gene to the gene for GFP, fluorescence can be used as the signal of successful transfer. Researchers have since mutated GFP into different forms that emit light in a variety of different colors, not just green. I would imagine that the essential ingredient in an Elvish lantern would be a culture of bacteria or fungus (such as yeast) in which a gene for GFP, or something like it, has become incorporated.

Something similar might explain how Frodo's sword, Sting, emits blue light in the presence of Orcs. Many bioluminescent organisms or photoreactive substances emit light only when triggered by some external event, usually manifested as the presence of some activating substance. It could be that the material from which Sting is made contains a substance that acts as a chemical sensor — an artificial 'nose' — specifically tuned to Orkish exhalations. Reacting to this, the sensor could then trigger some form of GFP that would emit blue light, alerting Frodo to the presence of Orcs.

Bioluminescence provides a rather obvious example of how Elvish technology would take its cue from nature, offering a characteristically simple solution to a problem that is at the same time organic, very neat and, at first sight, magical — there are no wires, bulbs, or batteries to spoil the effect.

A less obvious sign of Elvish technology is a characteristic tendency towards modifications of the perception of *time*. This might seem rather a stretch, but it is, in fact, perfectly in accord with the Elvish outlook on the world. As creatures that are immensely long-lived and yet bound to the fate of the Earth, which has a finite if very long life, the Elves are sensitive both to time and our experience of it — and they can, by virtue of having a lot of time to play with — experiment with it.

This suggestion is at first sight outrageous, but I offer as my first piece of evidence the text itself: in Lothlórien, the most 'Elvish' of all the locales in *The Lord of the Rings*, the Fellowship experiences apparent distortions in their perceptions of the passage of time, as if time itself had stood still: this, it turns out, results from the preservative effects of Galadriel's hidden Ring,[10] explicitly a product of Elvish technology. Similar effects are evident in Rivendell, as Bilbo notes to Frodo.[11]

As my second piece of evidence I advance the star-glass, the phial that Galadriel gave Frodo as her parting gift. The phial contains the light of Eärendil's Star (that is, Venus) trapped in water, but not just *any* water — specifically, the waters of her fountain (*Rings* II,8). The star-glass would emit this light apparently of its own accord, or if prompted should its bearer find himself in dark places. To do this, the star glass must somehow store the light, emitting it somewhat later: because our perception of time depends on the speed of light, it is plain that any alteration in the way that light passes through substances amounts to a manipulation of time. Now, it could be that the phial, the water, or a combination of both, act as a kind of holographic camera, storing the information for later retrieval, in a similar way to the *palantíri* I described in 'The Laboratory of Fëanor.' This does not ring true, however, because the star-glass is acting quite specifically as a repository of *light* whose effects are present and immediate, not *images* that record an event from the past. I'd like to suggest instead that the water is, or contains, an exotic state of matter that slows the passage of light through it by a significant degree.

The speed of light is 29,979,245,800 centimeters per second, but only in a vacuum. When passing through anything more substantial, light slows down, and the effect of this retardation is to warp a beam of light, a phenomenon known as 'refraction.' When we see someone passing a window, we do not see them as they are *now*, but as they were a fraction of a second in the past: the light reflected from their bodies has had to traverse the air between us, and the glass in the window, and this journey takes a finite amount of time, even if too small for us to notice.

The Light of Other Days, an evocative short story by the late Bob Shaw, is based on the idea that the time it takes light to pass through a pane of glass might be dramatically increased, so that the image we see through a window need not be the present reality, but a scene impressed on the pane months or years earlier. In the story, a honeymooning couple on a motoring tour stops to investigate the wares of a roadside dealer in 'slow glass.' The dealer is exposing several large panes to the country scenery – once transported and installed in a city, you would see, instead of the dull urban prospect, the same rural view, in complete reality, for many years, the lifetime of the glass dependent on the speed at which images pass through it, and the time when the glass was last exposed to the view in question. The couple happen to glance through the windows of the dealer's cottage and see images of blissful domesticity – when they enter, the cottage is revealed as a bare shack, the homely pictures simply a memory of a lost life, still emanating from the slow-glass windows.

Since Shaw's story was published, research on various extreme states of matter has revealed how light passing through material can be slowed down very dramatically, or even stopped altogether.[12] The work depends on quantum-mechanical effects in which pulses of light can be persuaded, in certain circumstances, to propagate through an otherwise opaque soup of ultra cold atoms.

I propose that Galadriel's star-glass works very much like Shaw's 'slow glass,' except that the active material is the water, not its crystal container. The water in Galadriel's fountain is exposed to the light of Eärendil's star, which is then trapped in the liquid, and which is released very slowly after being bottled. That the star-glass is working by the retardation of light, rather than simply replaying a hologram, comes from the revelation that its life is finite, and that it ceases to function once Frodo and Sam had reached the Sammath Naur, the Cracks of Doom in Mordor.[13]

At that point the very last light to have impinged on the water before it was bottled had finally left the medium: given what we know of the chronology of the story (*Rings* B), this clue allows us to work out how fast light travels in this medium.

Galadriel gave Frodo the star-glass on February 16, so this represents the very latest time that the glass can have been prepared. The Hobbits notice that the glass has lost all potency on March 25, some 39 days later (given that February has 30 days in the Shire calendar: *Rings* D), and we can assume that this represents the very latest instant at which the glass expired. Let us assume that the star-glass is a centimeter in diameter: this means that the speed of light through the star glass is $1/39 = 0.026$ centimeters per day. Given that the speed of light in a vacuum, calculated from the speed above, is approximately 2.6×10^{15} cm/day, this gives a retardation factor of approximately $(2.6 \times 10^{15})/0.026 = 9.9 \times 10^{16}$ or about 17 orders of magnitude. In layman's terms, light traveling through the star glass is slower than its speed in a vacuum by a factor of 100,000,000,000,000,000, or 100 quadrillion.

This result is, in fact, an average value, for it assumes that the light has moved through the medium at a constant rate without either slowing down or speeding up. However, experiments on the passage of light through preparations of ultracold atoms show that the speed can be modulated by external influences such as lasers. In the case of the star-glass, emanations can be influenced by sonic vibrations 'tuned' to the medium: the light is brighter when Frodo or Sam — in response, say, to confronting Shelob — voices an encouraging exclamation in Elvish (*Rings* IV,9), suggesting that the voice prompts a more rapid passage of light through the medium, with the consequence that its overall life will be shortened.

Another reason for proposing that the active medium is the water inside the phial, not the phial itself, is that the Mirror of Galadriel may share some of the same properties. The Mirror is made of water in a bowl, the water emanating from the same source as that used to create the star-glass. The water in the Mirror clearly has strange optical properties, but I confess to being somewhat at a loss to explain them, except to say that they clearly relate to the ways and means in which Elvish technology distorts the perception of time. It could be, however, that much of the effects of the Mirror depend on autosuggestion, given that no two people looking at the Mirror see the same images. It may also be that these time-shifting properties are related to the Silverlode itself, presumably from the same watershed as the source of the water in Galadriel's fountain. When Gimli, Frodo, and Sam look into the lake of Kheled-zâram (*Rings* II,6), at the headwaters of the Silverlode, they see the glinting stars of the Elder Days before the Sun and Moon, even though it is full daylight, but they cannot see their own shadows or reflections.[14]

There is another possibility, though, in which temporal distortion is pushed to its limits: that the Mirror of Galadriel opens up a kind of wormhole, offering the viewer glimpses – subjective, and hard to control – of alternate universes. It may be no coincidence that this happens in Lothlórien whose connection with the general stream of time seems in general somewhat loose. It could be that in the Mirror of Galadriel we have a singularity, in which a part of Lothlórien loses its connection with time completely.

18. IN THE MATTER OF ROOTS

If there is one thing that gets Tolkien fans in a stew (apart from whether have Balrogs have wings) it is the question of the provenance of certain plants in Middle-earth that really shouldn't be there: plants such as tomatoes, potatoes and tobacco. As is so often the case, the problems that preoccupy fans are the easiest to solve. For example, staff at TheOneRing.net have told me that the question they get asked most often is the origin of Hobbits, when Tolkien states quite explicitly that they are close kin to Men — this in the Prologue to *The Lord of the Rings*, so one doesn't have to dig very deep. The hard problems (which are, of course, the most interesting) are considered less often.

However, every Tolkien reader knows and appreciates that behind the simple foreground lie progressively more remote and undiscovered vistas that lend enchantment to the story. In this vein, I'd like to show in this chapter how relatively simple answers to questions about Middle-earth's botany illuminate scientific problems about the distribution of plants as a function of climate change, human intervention, and the passage of time, and also how Tolkien used botany as an evocative literary device in *The Lord of the Rings*.

The flora of Middle-earth as met in *The Lord of the Rings* can be divided into five main components (not all mutually exclusive) and I shall discuss each of these in turn.

The first and largest component is what I shall call the 'English hedgerow' flora, consisting of plants observed commonly in England in Tolkien's time.[1] Tolkien used these plants to lend color

193

and mood to a scene, which would be enhanced by their familiarity to the reader. Snap-dragons and nasturtians conjure up images of cottage gardens; oaks and beeches suggests lowland forest of the kind commonly found in northern Europe (and therefore the environs of the Shire), whereas reference to firs, birches, pines, gorse and heather suggests a more high-altitude landscape. From references in *The Lord of the Rings*, it is clear that all of Middle-earth north and west of the White Mountains, and east to Mirkwood and beyond, was dressed in these familiar clothes.

Considering how well-realized Middle-earth is in terms of its landscape, there are surprisingly few direct references to identifiable plants in what I have called the familiar English-hedgerow flora of Middle-earth: in what follows I include references to direct observations of plants in a wild or domestic setting, or as used in artifacts (such as yew wood in bows), but I exclude references to plants in poems or songs.

These botanical references include the snap-dragons (*Antirrhinum*), nasturtians (*Tropaeolum majus*), strawberries (*Fragaria* sp.) and plums (*Prunus* sp.) of Shire gardens (see *Letters* 148 on the use of 'nasturtian' rather than 'nasturtium'); the apples (*Malus*) of Shire Orchards (and the apple-wood of Tom Bombadil's hearth); forest trees such as alder (*Alnus*), ash (*Fraxinus*), beech (*Fagus*), birch (*Betula*), chestnut (*Castanea*), elm (*Ulmus*), fir (*Pseudotsuga*),[2] hazel (*Corylus*), hawthorn (*Crataegus*, including uses of 'thorn' in this specific sense, when hawthorn appears to be what is meant), holly (*Ilex*), linden (that is lime, *Tilia*), oak (*Quercus*), pine (*Pinus*), rowan (*Sorbus*), willow (*Salix*) and yew (*Taxus*), and miscellaneous plants including bilberry (*Vaccinium*), brambles (*Rubus*), hartstongue fern (*Phyllitis*), heather (*Erica* or *Calluna*), ivy (*Hedera*), sloe (also known as blackthorn, *Prunus*), water-lilies (*Nuphar* or *Nymphaea*),[3] whin (that is gorse, *Ulex*) and whortle-berry (*Vaccinium*, a variety of bilberry).

These references occur sporadically throughout the text, but there are two notable concentrations, in 'Three's Company' (*Rings* I,3) when the landscape of the Shire is described lovingly and with regret, just as the Hobbits are about to leave it; and in 'Treebeard' (*Rings* III,4) where a rash of trees is mentioned as personifications of Ents.

In this context, it is notable that of all general categories of plant, Tolkien mentions and describes trees in much the most detail. His love of trees is too well known to deserve specific documentation, however it may be worth mentioning that *The Lord of the Rings* mentions all but three genera of tree native to Britain. According to the Woodland Trust (a British charity devoted to the preservation of woodland), there are twenty-one genera of trees and shrubs native to Britain: of these, Tolkien appears to have omitted only the hornbeam (*Carpinus*), field maple (*Acer*) and the poplar or aspen (*Populus*). However, two of the native genera mentioned in the *Lord of the Rings* — *Juniperus* and *Buxus* — occur only in the 'southern' context of Ithilien, of which more later.

In contrast, Tolkien says very little about the shrubs and herbaceous plants one might have met in woods, clearings or by the roadside, as one might on a country walk in Britain. Umbellifers are mentioned rarely — the hemlocks (*Conium*) mentioned in the fragment of the *Lay of Leithan* that Aragorn sings to the Hobbits, and in a clearing in the Old Forest (*Rings* I,6; I,11): Tolkien has little place for — say — buttercups, daisies, and dandelions.

This aside, when you look at how Tolkien uses his flora as adornments to the mood of the story, as distinct from the mechanics of telling it, it becomes clear that one of the main reasons for discussing specific detail is to make the reader identify with the landscape as one immediately familiar to the reader, who will therefore feel 'at home' in it. For the Shire, the reason is obvious: we are meant to feel that this is somewhere that we, like the Hobbits, belong — and that everywhere else will seem doubly strange and dangerous as a consequence. As regards Fangorn forest, the welter of familiar trees dulls the initial feeling of strangeness, so we can empathize with the otherwise alien Ents.

In contrast, when Tolkien wishes to use landscape to conjure up feelings of mystery or threat, he is very vague about the botany. When the Hobbits enter the Old Forest (*Rings* I,6), for example, five pages go by before any single individual tree is isolated from the menacing collectivity (indeed, some of the trees are described as 'nameless') — a dramatic contrast with the earlier descriptions of the Shire. The Old Forest is sufficiently close to the Shire for us to be sure that the trees it contains will not be wildly exotic, so the failure to mention any by name must be part of the literary effect — making, as a result, the eventual meeting with Old Man Willow all the more dramatic. In another case, the stunted flora of the Morgai, the inner fences of Mordor, is described in the most general terms, as tussocks and thorns (*Rings* VI,2), summoning up a landscape of drear desolation. The Hobbits are not here to admire the flowers: on the other hand, Tolkien resists the opportunity to list a catalogue of cacti.

This opportunity for detail is, however, welcomed with open arms as regards the second of my five floral components of Middle-earth: the flora of Ithilien as described in the chapter *Of Herbs and Stewed Rabbit* (*Rings* IV,4). In the story, Frodo and Sam come to Ithilien after having clambered over the barren Emyn Muil, slogged their way through the Dead Marshes and toiled over the desert before the Black Gate. They had not seen any plant of note since they left the Fellowship behind at the Falls of Rauros, so their sudden encounter with the woods of Ithilien is meant to be an intense, almost overpowering experience. At the same time, we are being introduced to a markedly different floral province from the English-hedgerow variety seen up to that point (although some of the plants mentioned are, in fact, native to Britain — that is, the Shire — so the Hobbits would have recognized them). To this end, Tolkien overturns the sparseness of his earlier botanizing, and in the space of just two pages mentions more kinds of plant than in the rest of the story combined.

The plants mentioned in these pages are sufficient to show that the flora of Ithilien is meant to have a southern, somewhat Mediterranean character: sheltered from the east by the mountains of Mordor, Ithilien represents the northernmost extent of a flora perhaps seen (although never mentioned) along the coastlands of the Bay of Belfalas and to the south of Mordor. Trees not previously mentioned include bay (*Laurus*), cedar (*Cedrus*), cypress (*Cupressus*), larch (*Larix*), tamarisk (*Tamarix*), terebinth (*Pistacia*), olive (*Olea*) and unspecified 'resinous trees,' together with species seen earlier such as pine, fir, and filbert (that is, hazel). In marked contrast with the rest of the book, Tolkien's description of the shrubs and herbs of Ithilien outweighs that of the trees, and includes myrtle (*Myrica*), thyme (*Thymus*), sage (*Salvia*), ling (a general term for heather, mainly *Calluna vulgaris*), broom (*Genista*), cornel[4] (*Cornus*), juniper (*Juniperus*), marjoram (*Origanum*), parsley (*Petroselinum*), saxifrage (*Saxifraga*), stonecrop (*Sedum*), primerole (that is, primrose, *Primula* sp.), anemone (*Anemone*), asphodel (*Asphodeline*), lily (Liliaceae), water-lily (*Nymphaea* or *Nuphar*), iris (*Iris*), eglantine (*Rosa*) and clematis (*Clematis*) – as well as briars encouraged by the disturbance of Orcs, to indicate the encroachment of Mordor. Later chapters set in Ithilien (*Rings* IV,5; IV,7; VI,4) mention ilex (the holm – or evergreen oak, *Quercus ilex*), box (*Buxus*), celandine (*Chelidonium*), anemones, hyacinths (*Hyacinthus*), gorse, whortleberry, stonecrop and – in Gondor (*Rings* V,8) – the roses 'of Imloth Melui,' presumably derived from a wild species such as eglantine.

Tolkien is at pains to stress that the flora of Ithilien, as well as Gondor south of the White Mountains, includes plants and trees unfamiliar to Hobbits, even one so garden-learned as Sam, and which are indeed unknown in the Shire. These include the unspecified flowering creeper growing across the brow of the fallen king at the Crossroads (*Rings* IV,7); and the dark-leaved trees with scarlet flowers at the Field of Cormallen (*Rings* VI,4) which may or may not be related to the evergreens observed by Merry and Pippin in Fangorn, which had dark glossy leaves like thornless holly, and upright flower spikes bearing shiny olive-colored buds (*Rings* III,4). A curious example is the unidentifiable source of the hardwood *lebethron*, made from a tree found in Gondor, used to make the walking sticks given to the Hobbits by Faramir, as well as the casket containing the crown of the Kings (*Rings* IV,7; VI,5). An even odder case is the wood used to make the Elven boats gifted to the Fellowship as they left Lórien, of a nature obscure even to a woodland Elf such as Legolas (*Rings* II,9).

At this point, the unfamiliar flora of Ithilien grades into the third category of the flora of Middle-earth – plants characteristic of Middle-earth with no obvious counterpart in the modern world, possibly because they have become extinct, along with Ents. These plants include the 'evermind' or *simbelmÿne* on the barrows of the Kings of Rohan, the *elanor* and *niphredil* of Lórien,[5] the semi-mythical 'White Tree,' and that most Tolkienesque of trees, the *mallorn*.

This species occurs only in Lórien, where it appears to have an almost symbiotic relationship with the Elves. The botanical affiliation of this tree is hard to fathom, although the clues left by Tolkien suggest a close resemblance to the beech: like the beech, the *mallorn* propagates from a smallish nut, it is deciduous, and its leaves, after turning to gold, remain for a long time on the branches before being shed. As with beech, little appears to grow in the shade of the *mallorn* — there seems to be little in the way of shrubbery or understory in Caras Galadhon. I propose (necessarily speculatively) that *mallorn* is a variety of beech selected by the Elves for its girth, height, and characteristics of leaf fall. The great longevity of forest trees is the principal reason why people have not bred and selected them in the same manner as herbaceous plants. Were that not the case, oaks with edible acorns would be a part of our diet.[6] Elves, however, live far longer than trees as well as people, so there should be no reason why Elves should not have bred trees in the way that humans breed cacti, orchids, or any other short-lived plant.

The fourth category of Middle-earth's flora includes those domesticates that appear to have come from somewhere else. These include potatoes, tomatoes, tobacco, tea, and sunflowers.

From the first, potatoes are shown to be a staple food crop of the Hobbits. Sam's father, the Gaffer, is acknowledged as an expert in their cultivation (*Rings* I,1), and Samwise longs for a few 'taters' to boil in Ithilien (*Rings* IV,4)[7] — some chips to fry with Gollum's fish. Tomatoes feature in Peter Jackson's film *The Fellowship of the Ring* as a regular ingredient of Hobbit campfire fry-ups. Tolkien mentioned them in the first chapter of an early edition of *The Hobbit*, but later substituted 'pickles' in their place.

Tobacco, however, is featured most prominently and discussed extensively by Tolkien, either in his own voice or through that of Meriadoc (*Rings*, Prologue). It is plain that tobacco is, as in the real world, a variety of *Nicotiana*, grown in Gondor for the scent of the flowers; which flourishes in sheltered spots in the South of the Shire; and was originally brought across the sea by the Numenoreans. This single clue, right at the beginning of the *Lord of the Rings*, solves the problem, and which really ought to have cleared up the disquiet of Tolkien fans. For if tobacco can be brought from across the sea, so can anything else.

Nicotiana tabacum, potatoes (*Solanum tuberosum*) ,and tomatoes (*Solanum lycopersicum*) all belong to the great plant family Solanaceae, whose members are found all over the world.[8] Examples that one might expect to find in some sorcerous brew in Middle-earth are mandrakes (*Mandragora*); nightshades, too (*Atropa bella-donna*), would presumably have featured in Shire hedgerows. However, all the members of the Solanaceae that we associate with food — eggplants (*Solanum melongena*) and chili peppers (*Capsicum*) as well as potatoes, tomatoes, and tobacco — are indigenous to Central and South America, and were introduced into Europe around the sixteenth century, after the first European voyagers returned from the New World.

Other exotics, mentioned more fleetingly, are sunflowers and tea (both in *Rings* I,1). The sunflower (*Helianthus*) is a native of the New World, though it belongs not to the Solanaceae but to the Asteraceae (which also includes daisies and Jerusalem artichokes.) Sunflowers could have been brought to Middle-earth by the Numenoreans along with tobacco and potatoes. Tea (*Camellia sinensis*) may pose a more persistent problem. The Hobbit habit of tea-drinking was all part of the smug Englishness that Tolkien wished to convey. However, tea-drinking depends on reliable and constant imports of tea, originally from the Far East. For Hobbits to have drunk tea as a regular staple rather than as an occasional luxury item, they would have been more conversant with trade and the outside world than they evidently were.

Given this context, one might ask why Tolkien didn't have his Hobbits subsisting on an Old English diet of turnips and drinking infusions of dandelion leaves. However, Hobbit society is in some ways anachronistically modern (exemplified by Bilbo's silk waist-coats and Lobelia Sackville-Baggins's umbrella), so much so that potatoes and even tea seem commonplace, even to the most unadventurous of Hobbits, among whom the Gaffer stands as an almost parodic example. And yet, given the propensity for the Numenoreans of the Second Age to explore every inch of the available world (See *Akallabêth* in *The Silmarillion*, and *The Mariner's Wife* in *Unfinished Tales*), one might have expected them to have brought a few plants back from their explorations, some of which would have become naturalized in Middle-earth.

If any further proof were needed that some of the plants of Middle-earth came from somewhere else, we should look no further than the aromatic herb kingsfoil or *athelas* (for which there is no clear real-world equivalent). This is mentioned quite explicitly (*Rings* I,12) as a plant introduced by the Men of the West, now found naturalized in places where they once dwelt or camped.

The fifth and final category includes flora of the Shire that seem singularly inappropriate for reasons of climate. One might stretch a point to include *Nicotiana tabacum* in this category (given that *Nicotiana* is easily grown in England as an ornamental), but I refer in particular to the domestic grape *Vitis vinifera*. Now, grapes have been grown for winemaking in England and Wales since the Conquest, and possibly even in Roman times. Although the damp climate has made the practice a somewhat uncertain enterprise, wine-making is currently thriving in England and Wales. More than ninety per cent of English and Welsh wine is white, made from hardy German grapes: only a tiny amount of English or Welsh wine is red. (By the way, I refer to 'English and Welsh' quite deliberately: Scotland is too far north for winemaking, and the term 'British wine' refers specifically to wine made in Britain from imported grapes.)

Secure in this knowledge, one can only raise an eyebrow at Bilbo's mention (*Rings* I,1) of Old Winyards, a strong red wine from the Southfarthing. Vines had been grown in the Shire since the time of the Kings (*Rings* Prologue), and what with evidence for much long-term climate change, the association of wine and tobacco with the Southfarthing, and barley and snow with the Northfarthing, suggests a dramatic climatic contrast across a country that measures just fifty leagues – about 150 miles – from North to South, equivalent to the distance between Leeds and Oxford, or slightly less than the distance between Manhattan and Boston.

I have three ways to explain this dramatic contrast between the Shire and the England on which it is otherwise so closely modeled. The simplest is to propose that the Green Hill Country, running east to west on the northern edge of the Southfarthing, provides a convenient shelter from northerly breezes, dividing the Shire into distinct northern and southern floristic sub-provinces. The second explanation is also one of geography: that is, the Shire, unlike England, is landlocked, and presumably far from the maritime influences that make the English climate so mild and wet (and which conspire to make red wine hard to make).

The third way is all about climate. The climate of England has varied over historical time, and the climate of the Shire at the end of the Third Age is reported as relatively mild. The effects of climate change on botany can be surprising, and very marked over long periods, producing flora that might seem unexpected were the climate to remain constant.

Over the past thousand years, the climate of Britain has included prolonged periods of warmth and cold; over the past million, it has varied from tropical to almost polar. England contains few native reptiles today, and no turtles, but a few hundreds of thousands of years ago, a species of tortoise called *Emys orbicularis* bred in the British Isles.[9] This is notable because this creature requires summer temperatures in excess of 18°C for its eggs to hatch. Hippos once wallowed as far north as the River Tees, and lions stalked oxen in what is now Trafalgar Square in London. Exotic plants such as the water-chestnut (*Trapa natans*) could be found in English streams. Roll the tape of time forward to fifty thousand years ago, and England was a nearly treeless steppe supporting herds of reindeer in winter and bison in summer.

The effects of rapid climate change over the past million years or so has been to create some very unusual floras. The most contentious case concerns the so-called 'mammoth steppe' of northern Europe and Siberia, the predominant environment of the Ice Age.[10] This habitat included plants nowadays associated with the Arctic alongside those of a more southern character, and supported an enormous tonnage of mammoth, bison and other game hunted by our ancestors. Some scientists think that the mammoth steppe was a habitat of a kind not seen nowadays: others, however, have suggested that it never existed, and that the various disparate elements were thrown together randomly as a result of the vagaries of fossilization. The consensus is swinging to the first of these two views, but the debate highlights the problems faced by scientists when trying to make sense of the unknown in terms of what is familiar. In any case, it is only fair that even though the Shire is based on England, it should contain a hint of exoticism, After all, England no longer contains Hobbits or talking trees.

More fancifully, the reference to Old Winyards may, like so much of Tolkien's work, be weighted with nostalgia and regret: it recalls Keats' melancholic longing for oblivion in some sun-drenched refreshment in *Ode to a Nightingale*:

> O for a draught of vintage! that hath been
> Cool'd a long age in the deep-delvèd earth,
> Tasting of Flora and the country-green,
> Dance, and Provençal song, and sunburnt mirth!
> O for a beaker full of the warm South!

It may also be worth noting that in the summer of 1940, as Tolkien was thinking about and drafting the opening chapters of *The Lord of the Rings*, nostalgia would have been heightened by the strong possibility that England faced imminent annihilation. There were desperate dogfights between Spitfires and Messerschmitts against the clear blue background: that Battle of Britain summer of 1940 being one of the warmest and driest on record.

19. THE ONE RING

We come to it at last: that small, seemingly insignificant artifact on which rests the fate of Middle-earth. I have put off discussing the One Ring because I cannot, as yet, propose a fully convincing explanation for how it works.

This does not mean that I haven't had a jolly good try, and many of my friends and colleagues have added their own voices to what is something of a collective effort. So what you see in this chapter is a range of partial solutions, most of which overstep the boundary between real physics and complete speculation: however, some of them might set you on the right path, or spark some completely new line of inquiry. It could be that there is a very obvious solution, under our very noses, but we have not had the wit to see it — maybe some chance and completely unrelated happening will jolt our minds in the right direction. This questing, this uncertainty, is nothing to be ashamed of. Were we always to have a definitive answer, science would stop. Science is a road that runs forever on, whose answers are always provisional, constantly subject to revision, even complete reversal.

Let us try to summarize the history of the Ring as a way to approach its properties. It was made by Sauron in the Second Age as a way to unmask and control the Three Rings of the Elves. When Sauron first put it on, his thoughts were revealed to Celebrimbor, maker of the Three, after which the Elves kept their Rings hidden (*Rings* II,2). To this extent, the Ring is rather like a *palantír*, and one might suppose a degree of quantum entanglement between the Three and the One, similar to that which I proposed existed between the Seven Stones in 'The Laboratory of Fëanor.'

Second, Sauron used the Ring as a device for storing aspects of his own personality and power: without it, he cannot achieve full domination of Middle-earth (*Rings* I,2). In one way, the Ring joins the Two Trees and the Silmarils as objects in which their creators vested so much of their craft that they could not repeat the feat. There is another consequence of this, however, than the fact that were the Ring to be destroyed, Sauron would be unable to make another one. That is, that the Ring itself appears as a character, almost as clever and deceitful as Sauron himself, although it is not as fully autonomous as — say — Frodo, or Gandalf. In technological terms, the Ring can be seen as a storage device of extremely high density, sufficient to capture some aspects of the 'personality' of its maker, perhaps just enough to exert its own 'will' in particular situations.

For it to work properly, the Ring can only be worn by Sauron. When worn by others, the resulting mismatch (an electrical engineer might call it an 'impedance mismatch') leads to a range of pathological and psychological effects. Sauron is immortal, so the effect of wearing the Ring on mortals will be to prolong their lives beyond their normal range, as evidenced by Gollum and Bilbo. The wearer is additionally infused with an inappropriate lust for domination, each according to their stature: Gandalf or Galadriel would dominate the Earth, Boromir only Gondor, and Samwise a Garden, albeit expanded to the size of a kingdom: in other words, those under the influence of the Ring adopt as much of Sauron's own thirst for domination that their minds can conceive. However, in tune with Sauron's character, especially before the end of the Second Age, as a fair-seeming flatterer, achieving his will by suggestion rather than outright force, the Ring exerts its power by pumping up the thoughts of the wearer rather than by imposition from without.

Another feature of the Ring is that it changes size and weight apparently of its own accord. It seems to require constant attention, can slip from a finger all on its own, and can appear unexpectedly heavy. This apparently autochthonous behavior chimes with the idea of the Ring as a character; it also suggests that it induces a state of preoccupation, even obsession, in the wearer, a condition that Tom Shippey has likened to drug addiction. However, the changes in size and mass of the Ring might be physical as well as psychological, and there might be a way to account for the physical and the psychological effects of the Ring in a single, theoretical framework.

Nobody knows how much capacity would be required to store a human personality offline, if indeed such a thing is possible. A mind is, after all, not a 'thing,' but a summation, a manifestation of the workings of the body and brain. It works in different ways in different circumstances, and at times even winks out altogether without our noticing.[1] Tolkien's conception, at least in Middle-earth, was profoundly different, in that the body was seen as a 'house' (hröa) for the distinct and separate entity of the spirit (fëa).[2] Nevertheless, Sauron's personality — spirit, essence, mind — would have been of such greatness that one might wonder how he could store any significant part of it in a small golden ring.

One solution might be that the Ring is not an object, in and of itself complete, but simply one facet or a greater object that exists in dimensions ordinarily inaccessible. By this, I do not mean that the Ring is a kind of 'gateway' to other worlds, but retains within itself a way to access dimensions that some physicists believe exist, yet which we are unable to experience. Sauron, being a spirit whose origin lay beyond the world we inhabit, might have worked out a way to store his personality in these extra dimensions.

For many years, physicists have had the idea that the three dimensions of space that we experience in our everyday lives — the dimensions of up-and-down, left-to-right, and back-and-forth — are insufficient to explain the interaction between matter and energy.[3] This unquiet spirit, often lurking in the background, has come to the fore in the effort to 'unify' the disparate forces of nature into one great 'Theory of Everything.'

There are four fundamental forces of nature, and they are, in order of decreasing strength, the 'strong' nuclear force, which is strong enough to keep atomic nuclei together; the 'weak' force, responsible for other interactions on the scale of atoms; the electromagnetic force, responsible for, among other things, the chemistry that makes people possible; and gravity, which governs the formation and interaction of planets, stars and galaxies. The scale at which these forces act is inversely proportional to their strength. The strong nuclear force can keep atoms from flying apart, but it becomes weak at any distance greater than the nuclear. The electromagnetic force is much weaker, but we experience it in everything we do. Gravity predominates on the planetary and cosmic scale but is very weak over small distances and with masses smaller than that of a minor planet: the gravitational force between everyday objects, such as two apples, is virtually unmeasurable. The only reason why apples unerringly fall to Earth from trees, as reputedly observed by Newton, is because the Earth is huge, and, more pertinently, very much more massive than an apple.

Recently, physicists have managed to accommodate electro-magnetism and the 'weak' and 'strong' nuclear forces in the same theoretical framework, but nobody has yet succeeded in bringing gravity into the fold of unification. Ever since Einstein, physicists have tried to find a way to express gravity in terms of quantum mechanics — the physics of very small things, which often seems bizarre to those of us accustomed to living our lives in the realm of the macroscopic. Their efforts, however, have so far failed.

A peculiarity of gravity, as yet unexplained, is not just that it is weak but very much weaker than it ought to be, given our current understanding of the behavior of matter and energy. In trying to account for this, physicists have come across a promising line of enquiry in the theory of 'strings,' in which the fundamental units of matter and energy are not conceived as point-like particles, but extended, one-dimensional objects – the 'strings' of the theory. Crudely speaking, there is really only one kind of string: it is the way the string 'vibrates' that determines the character of each individual particle. This vibration is, however, metaphorical, because it refers to a process that occurs in 10 or 11 dimensions. String theorists assume that the three dimensions we experience are a subset of reality, but the remaining dimensions are somehow curled up very small (the technical term is 'compactified'). In string theory, gravity is just one more aspect of vibration, so to this extent string theory has been successful at unifying the forces of nature.

All, that is, for two things. First, the theory does not explain why, in our reality, we experience just three spatial dimensions (rather than any other number, such as two, or five, or ten), and why all the others are compactified. Second, the extra dimensions are curled up so small — many orders of magnitude smaller than atomic nuclei – that we have no hope of ever testing the validity of string theory.

Physicists are nothing if not endlessly ingenious, and have come up with two loopholes. The first is a model in which reality is not conceived in terms of one-dimensional strings, but two-dimensional membranes, or 'branes.' The second exploits this idea by supposing that everything we experience, our plane of reality, occurs on a 'brane' embedded in a higher reality to which only gravity has access. All the forces of nature except gravity are confined to the brane of reality, but gravity can leak away, explaining why it appears abnormally weak — it is shared out among one or more extra dimensions. These dimensions might be much larger than the tiny ones of classical string theory — of the order of millimeters, or more – and so might be amenable to testing, even experimental exploration. Because our senses depend on electromagnetism, which is confined to the plane of the brane, we will be unaware of the existence of these dimensions, which operate in parallel to our regular everyday reality.

I suggest that the Ring is a device that somehow has access to the extra dimensions normally penetrable only to gravity. Sauron secretes his personality in these extra dimensions, perhaps as gravitational wave information. The association with gravity may explain some of the other anomalies associated with the Ring — that is, its propensity to change size and weight, suggesting that it can transfer some of its mass (that quality of matter most associated with gravity) into these hidden dimensions, and back again.

The idea that the extra dimensions are of a size comparable to the ones we experience, and are not compactified to a subatomic degree as implied by string theory, offers an explanation for the presence of the world as experienced by the Ringwraiths. The long-term effect on mortals of wearing a Ring of Power is to gradually transfer their mass, personality, and even their being to these hidden gravity-dominated dimensions, the hideous world of the Nazgûl, who cannot detect electromagnetic radiation (which is why they are blind), but can nevertheless respond to the presence of matter, and are drawn in particular to the Ring.[4]

If that's all there was to the Ring, the problem would be solved (leaving aside, of course, the still-unproven existence of 'branes, or of other dimensions, let alone whether you could move between these hidden dimensions and our own, or store complex information gravitonically, as opposed to electronically.) Unfortunately, what really poses problems for the Ring is its most obvious property: that it makes its wearer invisible, along with his clothes and anything else he is carrying.[5]

Given the above, it might be supposed that the Ring imposes invisibility by transferring its wearer into gravity-dominated dimensions orthogonal to the 'brane. Each time the user puts on the Ring, a little of his mass is secreted into these hidden dimensions, explaining why a long-term user will feel 'stretched' and begin to fade, until he enters the hidden dimensions completely and becomes a wraith. In the short term, at least, the Ring confers invisibility by tripping a kind of switch, which happens when the Ring is penetrated by living matter. It could be that the Ring supports a perpetual supercurrent — that is, an electric current that is conducted without resistance, and which therefore has the potential to run unimpeded, and without dissipation, for very long periods of time. The presence of a finger inside the Ring influences various parameters of the supercurrent, activating some kind of switch that propels the wearer into another dimension.

Because electromagnetic radiation is confined to the 'brane of reality, the wearer becomes invisible, although he can still interact with the world. The wearer, once invisible, also experiences the world differently, in terms of gravitational rather than electromagnetic information. There is, however, a significant problem with this. That is, the wearer will not be able to 'hear' or 'see' in the normal way, because hearing and sight are conveyed by electromagnetic radiation. The fact that Bilbo (in *The Hobbit*) can hear and see Gollum, the Dwarves, the Wood-Elves, and so on while wearing the Ring renders the whole idea of a gravity-dominated dimension unsafe.

There are, however, other ways of becoming invisible. Recent technological advances have allowed scientists to create materials whose optical properties can be altered at the flick of a switch: for example, transparent materials can be transformed into mirrors.[6] The electrically induced change of optical properties is now something we take for granted. The liquid-crystal displays (LCDs) seen in the meanest wristwatch work because an electric current provokes a stack of organic crystals to change their physical orientation, which in turn changes their optical properties, making information visible. Could it be that the Ring produces some analogous change in the physical properties of the wearer? This would seem to be extremely unlikely, as such changes would render him blind (by making the eyes transparent), even if they do not produce changes in body chemistry that would kill him.

Another inexplicable property of the Ring is that the wearer can still cast real shadows, even if these are only very faint and detectable only if the wearer stands in full sun. This would be impossible were the wearer made physically transparent or transferred to another dimension. However, it might be possible were the wearer suddenly shifted a millisecond or so into the future, but remained in our own time-frame just long enough (oscillating back for a microsecond out of every millisecond, say) to cast a shadow. Time-shifting into the future has the advantage that the wearer could still interact with reality in a more-or-less normal way: the problem with this idea — apart from seeming unusually contrived, and the fact that we know of no way in which such time-travel might be effected — is that it might induce all sorts of relativistic problems that would interfere with the wearer's experience of reality.

The main problem with all these schemes, however, is the fact that it is not just the body of the wearer that is invisible, but all his clothes and effects. The eponymous hero of H. G. Wells' *The Invisible Man* did not have this problem, and wore visible clothes. The Ringwraiths wear visible clothes (over invisible robes, crowns and so on) but when Isildur, Gollum, Frodo or Bilbo put on the Ring, *everything* they are wearing becomes invisible.[7] I cannot explain how the Ring might exert such arbitrary effects. Either only the body of the wearer should become invisible, or the whole world, but not some arbitrary set of objects in between. Such a proposition makes any explanation needlessly complicated, and unnecessary complexity is always a bad sign for a scientific theory.

One way round this is to suggest that the invisibility is psychological rather than physical. The author Philip Pullman has come up with just such a mechanism. In *The Subtle Knife* and *The Amber Spyglass*, Will Parry becomes invisible by dressing and acting in such a bland way that he becomes totally inconspicuous. Woody

Allen exploited this idea to hilarious effect in his film *Zelig*, about a human chameleon who could blend into every social situation. This, presumably, is how spies operate – there is no physical reason why we cannot see such people, but they act in such a way that we do not notice they are there.

More deeply, this phenomenon exploits the fact that the external world is not arranged precisely as it is reported to our minds through the filters of our eyes and brain, which tells us what we want to see, rather than what is there. Conjuring and optical illusions have an ancient and venerable history of exploiting just this discrepancy, even in healthy people with normal capacities of sight and reason. The discrepancy can be widened, however, in certain disease states such as 'blindsight,' in which patients swear that they cannot see an object placed in front of them, even though their hands move unerringly towards the object in question.

Could the Ring somehow produce a switchable blindsight 'field' in anyone who might possibly look in the direction of the Ringbearer? It is a nice idea, and could explain the presence of shadows cast by the Ringbearer, but unfortunately suffers from a multiplicity of problems. First, it is difficult, and perhaps impossible, to imagine how the Ring, once worn, could induce a state of mass delusion among witnesses. Second, the mechanism leaves unexplained the whole phenomenon of the world of the wraiths.

The problems of the Ring stem from its literary history. In *The Hobbit*, it was introduced as a handy device that allowed Bilbo to fulfill his destiny as the best burglar this side of the Lonely Mountain. The sole purpose of the Ring in *The Hobbit* is to act as a cloak of invisibility, and to this end it is strictly a fairy-tale concept, a *deus ex machina*, of the order of Cinderella's glass slippers, or the amazing ability of fairy godmothers to turn pumpkins into golden carriages. As such, the convention – as with any conjuring trick — is that we do not inquire how the trick is done. When Tolkien was drafting *The Lord of the Rings*, however, he needed to engineer some link between his new story and *The Hobbit*, and the Ring was the obvious point of connection. *The Lord of the Rings*, however, grew away from the simple sequel that Tolkien originally intended, becoming vaster, darker, and much more 'realistic' in tone. As the story grew, so did the Ring, acquiring all kinds of sinister properties not evident in the earlier story, and which would not necessarily be compatible with that of 'simple' invisibility.

And so the case must rest. The fact that we cannot explain the Ring in terms of science or technology is not, however, reason for despair. To a scientist, the existence of the inexplicable is a challenge, and a reminder that science always has more to achieve – when we crest a hilltop after much effort, it is always a thrill to see rank upon rank of yet unconquered mountains of whose existence we had, until that point, been unaware. It is also fitting that there should be at least one thing in Middle-earth that remains inexplicable, and doubly appropriate that this artifact should be as iconic, as central to Tolkien's world, as the Ring itself.

20. SCIENCE AND FANTASY

Science is a word we use for the application of the fantastic.[1] All science that is enjoyable and worthwhile, rather than routine or directed in pursuit of some unconnected goal, starts when a person of vision looks outwards beyond the wall of what is known and asks the question 'What If...?.' What if we point our telescope at this galaxy rather than that? What if we mix these chemicals together? What would we see over the next hill?

The best science is driven by the curiosity we all felt as children, when virtually all the world was a strange new thing ready to be explored. When I was very small, my first scientific experiment was a test of the tolerance of small garden invertebrates to total immersion. Many professional scientists can recall similar childhood experiments — trying to measure the speed of fishes in aquaria, growing bacterial cultures, working out the distance of a car journey from the wavering needle of the speedometer when viewed from the back seat, or simply playing with matches. This curiosity is not limited by imagination or the concerns of others, particularly not by the needs of adults to — say — generate economic wealth or acquire power. It is simply a desire to explore new worlds, for the fun of finding out.

Science is questioning and subversive. Leaving experiments on drowning worms and beetles far behind, I found myself as a teenager in a science laboratory on the first day of a new term, with a new teacher, who said something which I now see was extremely brave. 'I may be the teacher,' he said, 'but you should never, *ever*, take anything I say as the truth. To be a scientist, you must be skeptical — you must test it for yourself.'

Science is license to imagine, to be creative, and to be free. Once this is accepted, it is very hard to see why a subject with such evident allure fails to attract more young people than it does. Instead, science is perceived as geeky, unattractive, unglamorous, smelly, and — well, too difficult. The number of people entering higher education is rising, but the proportion choosing to take science to college level is falling in both relative and absolute terms, in favor of subjects such as drama and media studies. Politicians and industrialists fret and worry about this decline, which they seem impotent to reverse.

The root of the problem, however, is not hard to find. It lies in the attitude of those scientists and politicians who have taken it upon themselves to address the question of the supposedly unwelcoming attitude of the public toward science — an attitude that is all about the triumph of authoritarianism over creativity, of imposed will over innate imagination.

Scientists have tried hard to stop the rot. The Royal Society, Britain's most senior scientific body, sought to address the question. To this end it convened a Committee for the Public Understanding of Science, charged with promoting science and raising its profile before a largely apathetic public. In tune with their program of awareness, various colleges set up courses in science communication, some of which are now well established and highly respected.

However, the good that these initiatives might achieve is marred by the attitudes of some of their more outspoken advocates. One of the best-known is the geneticist Richard Dawkins, who has written several critically acclaimed popular science books of which *The Selfish Gene* is perhaps the best known. My favorite, however, is *The Blind Watchmaker*, an articulate and passionate explanation of the mechanism of evolution, combined with a scorching critique of creationism. Dawkins' books offer a compelling message of the importance of science with both precision and power, and are without doubt classic works of popular science.

But Dawkins has not been content to let his books speak for themselves. He has taken a view that natural selection is a universal,[2] and that once that is accepted, all other explanations for human existence are untenable. I do not believe anyone with a scientific mind would disagree that natural selection is a force as mighty as Dawkins suggests. However, it does not follow from this that 'Darwinism' transcends science to become a belief-system that renders religion obsolete.

In any case, it is a tactical error for those promoting science to impose their view on others, for fear of replacing one kind of religion with another; diminishing the very subversive, questing quality that is the foundation of science; and, not least, for fear that the effort will backfire, turning people away from science rather than attracting them.

It is possible that people who had not previously thought that there might be a dichotomy between science and (say) astrology, such that they were obliged to choose one and forever abjure the other, might be inclined to ask why a scientist such as Dawkins is making such a fuss. If science is meant to be subversive, and not simply something whose nature is decided by authority, we — the public — should be entitled to take our own view, and that this view might not accord with that of Dawkins or anybody else. More dangerously, we might start to wonder whether science really *does* have any more intrinsic worth than the casting of runes, apart from the fact that its adherents shout more vociferously.

When the distinguished journalist John Diamond published his diaries of his own slow and grisly decline as a result of terminal throat cancer, the columns were collected in a book called *Snake Oil*, prefaced by an essay of the same name in which he criticized fellow sufferers who chose so-called complementary medicine over proven medical science. Nowhere, however, did Diamond insist that patients should be denied a choice to follow this path if they so wished, foolish though it might be — he simply pointed out, clearly and in detail, the statistics showing that complementary medicine for cancer is ineffective compared with conventional medicine. After all, Diamond attributed his disease to a longtime smoking habit, which he admitted was as much his choice as that of any particular therapy designed to remedy his illness. If patients die because of their own wrong-headedness, that is regrettable — but it is still their choice.

The effect is marred, however, by a preface from Dawkins endorsing Diamond's scientific standpoint and denouncing complementary medicine: and yet what is required is not stern reproof, but Diamond's method of quiet demonstration and example. People cannot be forced to make choices about things that matter to them by the imposition of the view of an external authority: they should be offered a choice, and then they will come to science by their own will, having learned that yes, science actually works, and there is no need to make a song and dance about it.

Worryingly, strident denunciations of activities perceived as pseudoscience will deter people from serious issues in which many would agree they really *should* be educated — but because people are turned off science, they will then not trust the sincerely meant advice of governments on issues that affect their lives, such as the siting of mobile-telephony masts, the growth of genetically modified crop plants, or whether to have their children vaccinated against diseases. The number of people open to purely scientific arguments (as opposed to those promoting some vested or political interest) can only be diminished by ignorance, especially when this ignorance is combined with a perception that science is connected with devious, secretive, and possibly authoritarian institutions. In such a climate, the arrival of a scientist exhorting everyone else how they should think will only make things worse.

As a means of making science palatable, this authoritarian stance is a clearly a failure: the more that its exponents rail and thunder, the more the public will turn away from science, and the more that young people will fail to engage with it. The failure lies in the fact that the proponents of scientific knowledge have lost touch with what it was like when they themselves first discovered science — that *frisson* when we first raised our heads above the cot rail to realize that there were undiscovered countries to explore, preferably when Mum and Dad weren't watching. Too many professors have failed to understand that imaginative acts stand at the very root of science — that any and every scientific investigation starts with a journey into the fantastic.[3]

As if to underline this poverty of imagination, the proponents of science have mounted direct assaults on the wellspring of science: fantasy itself. In August, 1998, the *Independent*, a British national daily newspaper, published an extraordinary denunciation of the fantastic by John Durant, Professor of the Public Understanding of Science at Imperial College, one of the leading research schools in Britain.[4] The occasion was the release in cinemas of a feature-length movie called *The X-Files*, based on the long-running and popular TV series of the same name. *The X-Files* is pseudoscientific nonsense, said Durant: we should turn away from such things and cleave to the truth of science.

In case you are unfamiliar with *The X-Files*, either the TV series or the movie, it concerns the adventures of two agents of the Federal Bureau of Investigation (FBI) charged with investigating paranormal incidents — alien visitations, mysterious disappearances, eldritch psychical phenomena, and so on. The show's success was due in large part to the differing attitudes of the agents to the mystifying cases they come up against in each episode. Agent Mulder (played by David Duchovny) is a True Believer, never shy to offer an unusual hypothesis to explain the data. Agent Scully (Gillian Anderson), in contrast, is a trained scientist and skeptic, whose specific task it is to shoot down Mulder's speculations, offering more rational and everyday explanations for the things they come across — a latter-day Dr Watson (if more glamorous). The key feature of *The X-Files* is that Mulder and Scully rarely reach any firm conclusions, usually for want of some vital piece of information. We, like they, are often left in the dark — the X-File remains open and the team is assigned another case.

It strikes me that Mulder and Scully are a perfect picture of scientific research collaboration, if perhaps more smartly dressed. Teams of scientists working the edges of the unknown are almost always in a position of having far too little information to explain the phenomena they see. Working and thinking always one jump ahead of the curve, the scientists propose hypotheses to explain the phenomena, which they then argue about, sometimes heatedly. Often unable to reach a conclusion, they devise experiments to test the various hypotheses they've come up with, and many of these experiments are inconclusive.

This messy and very human process of trial, testing, and frequent error is typical of scientific activity, yet something that spokesmen for science such as Dawkins or Durant tend not to articulate. The result is that when young people come across science, they may be shocked that it is not the effortless process of discovery and enlightenment that they might have expected, and so become discouraged. In addition, the publicizers of science rarely make clear that dispute and argument are usually signs of a science in robust health, rather than in a state of imminent dissolution — science is never, nor should it be, a united front, nor should all experiments be successful.

Argument and puzzlement are not, of course, things that professional communicators of science like very much. They like science stories to concentrate on undisputed facts, advertising great discoveries while minimizing the years of effort and false starts needed to get there. If you believe all the science stories you read in the papers, it is as if Professor Brainstorm went into the lab one morning, and eureka! Within an hour he finds the cure for cancer. After lunch he phones his local TV station and history is made. In reality, Professor Brainstorm has probably been working on a cure for cancer for years, trying all sorts of substances that don't work, thinking up all kinds of new and different experiments perhaps originally designed to cure other diseases, and having a few moments of inspiration and luck along the way. Professor Brainstorm's departmental managers would like him to deliver a five-year plan at which a cure for cancer will be delivered at the end. But real science doesn't work like that — it depends on creativity, questioning, and excursions into the unknown — an intuitive grasp of the fantastic.

What was striking about Durant's argument with *The X-Files* was that he failed to see that Mulder and Scully were attempting to address their admittedly outlandish situations in a perfectly valid and scientific way — a method driven by dispute, testing, and experiment, and informed by frequent failure. Instead, he directed his fire against the subjects of the agents' investigations - belief in alien visitations and paranormal experiences, as if the majority of the general public really found these credible. He argued that people should stop believing in such things, renounce them utterly, and subscribe to a list of known facts — which he and his learned colleagues would obligingly set out for us. Such a program is dangerous, and is but one step from Saruman's desire to set things neatly in order to his will, for what he patronizingly sees as the general good (*Rings* II,2).

Durant's argument assumes that most people treat *The X-Files* as a documentary series. One should not be deceived by the scientific trappings of the show, created with great care to enhance the verisimilitude of the setting, making the weird happenings stand out all the more prominently[5] — and the public is *not* deceived. Durant fails to understand that people are more intelligent than this, and that they engage with *The X-Files* as entertainment, placing it in the same category occupied by *Star Trek* and horoscopes. These things are fun — absorbing even — but few would believe for a minute that these things are objectively 'true.' As Tolkien made clear a long time ago in his essay *On Fairy Stories*, belief in the fantastic, in the context of a fairy story, requires the audience to suspend its disbelief for the duration, otherwise the effect will be spoiled. Once we have closed the book, we do not automatically believe that the world is full of Elves, Hobbits, and Dragons.

But if it so happens that some people do take astrology and paranormal phenomena seriously, why should they not? We live in a society that is as yet free. Conversely, scientists are not remote, white-coated drones who converse in algebraic formulae: they are real people like everyone else, people who go shopping, take their children to school, enjoy music and sport, vote in elections, fall in love, participate in organized religion and even read their horoscopes on the way to work (well, *I* do — don't *you*?). Scientists are people who have been drawn into science by the exercise of that very human capacity of imagining the unimaginable, not by learning facts by rote.

The same capacity for fantasy is found in doctors, lawyers, plumbers, accountants, soldiers, sailors and candlestick makers: and also the children from whose number the scientists of the future will be drawn. I might go further and say that scientists have retained a greater capacity for fantasy than anyone else, not less, and rather than condemning the act of fantasy as unrigorous, it is something that we should encourage young scientists to pursue with vigor.

Durant is wrong to worry that every person who watches *The X-Files* really believes in ectoplasmic emanations, aliens and flying saucers, because, as a rule, they do not. However, people enjoy imagining a world in which such things might be possible, and working out what their reactions might be were they to encounter them. As I noted at the start of this chapter, science is a formalization of that same capacity to place oneself in situations never before encountered. When confronted with an alien corpse, would *you* — the viewer — believe with Mulder that it is the real thing, or with Scully that further work needs to be done before we can be sure? The fact that the hard evidence invariably disappears before our intrepid pair can get their hands on it heightens the uncertainty — an uncertainty that is genuinely scientific, and which at the same time enhances our enjoyment of *The X-Files* as drama, not specifically connected with science.

Damnation of the paranormal should not also be a damnation of our ability to *imagine* the paranormal (irrespective of any scientific evidence for or against, or of whether we actually 'believe' in it or not, which is a separate question.) Nor should it be a damnation of our ability to conceive of anything outside our experience — an ability that would seem essential to discovery and the extension of knowledge, and something of which any proponent of science education would be expected to approve.

Were we all to take Durant's advice and believe only those things that he and his colleagues tell us are true (simply because they, the authorities, tell us so) science would be killed stone dead, instantly — and the tragedy would be that science educators would be genuinely puzzled at how science should have declined so catastrophically.

Science is not about the known, for that is boring. Science is about exploring the very limits of the unknown and trying to peer yet further into the gloom; to glimpse those vistas, dim and far away, to which Tolkien rightly attributed much of the charm of *The Lord of the Rings*.

In his essay *The Monsters and the Critics*, Tolkien argued that scholars of *Beowulf* had spent too much time excavating the ancient epic for clues about history, archaeology or linguistics, and not nearly enough time appreciating it as a work of literature in which the hero battles with a succession of monsters, unknown forces that stalked the wide and twilit world beyond Hrothgar's hall. As scientists, we must set our courses into the unknown. The monsters, said Tolkien, are what we should be looking at.

Ah yes, the monsters. Bring them on.

END-NOTES

Chapter 1: *Space, Time, and Tolkien*

[1] *Nature* 407, 137 (2000).
[2] See, for example, 'Experimental quantum teleportation' by D. Bouwmeester and colleagues (*Nature* 390, 575-579, 1997); this is the Innsbrück group mentioned by Simmons.
[3] This recurring nightmare appears in the words of Faramir during his wooing of Éowyn in the Houses of Healing (*Rings* VI,5).

Chapter 2: *Inside Language*

[1] I look at this problem in my book *In Search of Deep Time: Beyond the Fossil Record to a New History of Life* (New York: Free Press, 1999)..
[2] For more details on the evolution of vertebrate life on land see *Gaining Ground: The Origin and Early Evolution of Tetrapods* by Jennifer A. Clack (Indiana University Press, 2002) and *At The Water's Edge* by Carl Zimmer (New York: Free Press, 1999).
[3] 'Cladistic methods in textual, linguistic and phylogenetic analysis' by Norman I. Platnick and H.D. Cameron in *Systematic Zoology* 26, 380-385 (1977).
[4] 'The phylogeny of The Canterbury Tales' by Adrian C. Barbrook and colleagues, *Nature* 394, 839 (1998).
[5] 'Evolution, consequences and future of plant and animal domestication' by Jared Diamond, *Nature* 418, 700-707, 2002.
[6] Important papers on the link between genetics and the population movements of Neolithic farmers have come from Luca Cavalli-Sforza and colleagues (*Proceedings of the National Academy of Sciences of the USA*

85, 6002-6006, 1988) and Guido Barbujani and Robert Sokal (*Proceedings of the National Academy of Sciences of the USA* 87, 1816-1819, 1990.) Interestingly in this context, Tolkien noted that language would provide at least as good a test of ancestry as blood-groups (*Letters* 163), a contention that has since been supported by recent research: see Cavalli-Sforza, L. L., Piazza, A. & Menozzi, P. *The History and Geography of Human Genes* (Princeton Univ. Press, 1994).

[7] See 'Language-tree divergence times support the Anatolian theory of Indo-European origin,' by Russell D. Gray and Quentin D. Atkinson, *Nature* 426, 435-439 (2003), with an accompanying commentary by David B. Searls 'Trees of life and of language,' *Nature* 426, 391-392 (2003).

Chapter 3: *Linguistic Convergence*

[1] As a completely unrelated aside, Tolkien comments that Black Speech was itself an invented language, devised by Sauron for the use of his servants — and as such he failed in his aim, as all but his most senior captains used the Common Speech (*Rings* F). This could be an ironic and self-deprecating comment on the authors' own hubris as a subcreator of languages of his own, as well as an observation on the likely fate of all artificial languages such as Esperanto, invented with good intentions but which fail to thrive without an accompanying literature and mythology (see Ostadan's essay 'Glossopoeia for Fun and Proft' in *The People's Guide to J.R.R. Tolkien* for more on this topic). How could Orcs, a largely artificial people, ever cleave to an invented language, even one invented with them in mind?

[2] Iarwain Ben-adar, a Sindarin name of Bombadil (Rings II,2), looks suspiciously Hebraic in its construction. The word 'Ben' is the standard form denoting filial association ('son of'), and 'Adar' is a month in the Hebrew calendar that falls in early Spring and is particularly associated with jollity and merrymaking. The festival of Purim, which falls on the 14th Adar, commemorates the story of Esther, a tale reminiscent in its structure both of the *Arabian Nights* and also English mystery-play traditions — and is the only part of Hebrew liturgical tradition in which the name of God is not invoked. Although Tolkien intended Bombadil as a commemoration of the countryside of the Thames Valley, Bombadil's personification as a merry, capering fellow might suggest — albeit far

221

more speculatively — that Tolkien would have been aware of this aspect of Hebraic calendrical tradition. I am grateful to Jonathan Stein for suggesting this connection.

Chapter 4: *The Power of the Name*

[1] See, for example, A. J. Splatt and D. Weedon 'The urethral syndrome,' *British Medical Journal* Oct. 29, 1977, 1154.

[2] The conversation between Merry, Pippin, and Treebeard on the subject of species, names, and nomenclature (*Rings* III,4) can be read as an extended taxonomic joke, especially the part where Treebeard says that his true name in Old Entish is very long and growing all the time — and especially as Old Entish is represented as a kind of onomatopoeia in Quenya, or 'Elven-Latin' — a charge that could be easily laid at the feet of those seeking to name creatures in sometimes horribly bastardized versions of Latin and Greek. Another extended joke on taxonomy (or, perhaps, philology), this time rather more obvious and long-winded, is the by-play between Aragorn and the Herb Master in the Houses of Healing (*Rings* V,8) in which the Herb Master recites the many names of *athelas* before admitting that none of it exists in the House — and all this while patients lie dying and in need of this plant.

[3] Their guidance is formalized in the *International Code of Zoological Nomenclature*, 4th edition (Natural History Museum, London: International Trust for Zoological Nomenclature, 1999).

[4] See Clack, J. A. 'A new Early Carboniferous tetrapod with a *mélange* of crown-group characters,' *Nature* 394, 66 - 69 (1998).

[5] See Sampson, S. D., Carrano, M. T. and Forster, C. A., 'A bizarre predatory dinosaur from the Late Cretaceous of Madagascar,' *Nature* 409, 504 - 506 (2001).

[6] I am grateful to Mr. Mark Isaak and his fascinating website (http://home.earthlink.net/~misaak/taxonomy.html) reporting all the curiosities of biological nomenclature he can find, and for his permission to cite and discuss the names used herein.

[7] S. Conway Morris, 'Fossil priapulid worms,' *Special Papers in Paleontology* 20, 1-95 (1977).

[8] S. Conway Morris 'A new metazoan from the Cambrian Burgess Shale of British Columbia' *Paleontology* 20, 623-640 (1977).

[9] L. Van Valen, 'The beginning of the age of mammals,' *Evolutionary Theory* 4, 45-80 (1978).

[10] Anon., 'Naming the Loch Ness monster,' *Nature* 258, 466-468 (1975).

Chapter 5: *Holes in the Ground*

[1] See *Philosophical Transactions of the Royal Society of London, Series A*, 278, 447-509 (1975). I am most grateful to Turgon at www.theonering.net for alerting me to the Roberts reference, and trying in vain to convince me of a theory that the architecture of Bag End has more than a coincidental resemblance to the tunnels of the London Underground.

[2] See the *Rockall Times* (http://www.therockalltimes.co.uk/).

[3] An enjoyment that continues to this day (Roberts, personal communication.)

Chapter 6: *Inventing the Orcs*

[1] Lines 112-113 from *Beowulf in Hypertext* (http://www.humanities.mcmaster.ca/~beowulf/main.html), a resource created under the supervision of Dr. Anne Savage of McMaster University. The original Old English text and the translation are from the same source.

[2] Benjamin Thorpe, *The Anglo-Saxon Poems of Beowulf, The Scop or Gleeman's Tale, and The Fight at Finnesburh*. James Wright, Oxford University, 1865.

[3] I have not even begun to address the question of whether 'Orcs' are distinct from 'Goblins' either in degree or in kind; a matter on which one reader has taken me to task.

[4] *Rings* IV,9. There is a wonderful irony in Sauron's diminution of Shelob as his cat, in that Sauron's first appearance in the legendarium, in the earliest telling of the legend of Lúthien and Beren (*HOME* II), was as Morgoth's cat, even though one in gigantic form: Tevildo, Prince of Cats, living in his Castle, which is subordinate to Morgoth's greater fortress. In this early fairy-tale, Tevildo captures Beren and puts him to work in his kitchen. Lúthien rescues Beren by putting a spell on Tevildo and his gang of giant cats, reducing them to harmless kittens. This charming if rather contrived example of kitty lit, early abandoned, was a very far cry from Sauron in the cognate passage in *The Silmarillion*, as the werewolf in whose dungeons King Finrod and his companions — all save Beren —

are killed and eaten. Nevertheless, a trace of cattiness remains in Sauron: the lidless eye with a single dark slit — Sauron's representation in *The Lord of the Rings* — is much more feline than human, and is specifically described as being yellow, like the eye of a cat (*Rings* II,7). In the tower of Cirith Ungol, Shagrat and Gorbag, both Orcs, refer to the captured Frodo as a kitten, but Sam — still at large and dangerous — as a cat (*Rings* IV,10). Tolkien evidently had no love for cats, even in the real world. When asked by his publisher whether he would permit the use of names from his stories for a litter of Siamese kittens (*Letters* 219), Tolkien replied that, as far as he was concerned, cats belonged to the fauna of Mordor.

[5] D. M. Welch and M. Meselson, 'Evidence for the evolution of bdelloid rotifers without sexual reproduction or genetic exchange' *Science* 288, 1211-1215 (2000).

[6] L. G. Frank, M. L. Weldele and S. E. Glickman, 'Masculinization costs in hyenas,' *Nature* 377, 584-585 (1995).

Chapter 7: *Armies of Darkness*

[1] Isaac Asimov was an author of whose works Tolkien was especially fond (see *Letters* 294), even though he spelled his name 'Azimov.' If so, then Tolkien cannot have avoided Asimov's many robot stories, a number of which address questions about the extent to which the exhibition of human-like behavior in a machine qualifies the machine as human.

[2] See 'Knife-edge of Design' by Jack Cohen, *Nature* 411, 529 (2001).

[3] See *Letters* 153 and 269. In the first of these, Tolkien notes that Ilúvatar's toleration of the creation of Orcs from other beings is no worse theology than God's real-world toleration of the dehumanization of people by tyrants.

[4] See my essay 'Aspirational Thinking' (*Nature* 420, 611, 2002) and my book *In Search of Deep Time: Beyond the Fossil Record to a New History of Life* (New York: Free Press, 1999).

[5] The quote from Oken is from p.39 of Robert J. Richards *The Meaning of Evolution* (Chicago: University of Chicago Press, 1992), and is a translation from the original German, given in Lorenz Oken, *Abriss des Systems der Biologie* (Göttingen: Vandenhoek und Ruprecht, 1805).

[6] See my book *Jacob's Ladder: A History of the Human Genome* (New York, W. W. Norton, 2004) for more discussion of this point.

[7] E. S. Russell, *Form and Function: a Contribution to the History of Animal Morphology* (London: John Murray, 1916).

[8] Lamarckian strivings and man-from-the-apes fables of evolution are deeply buried in Tolkien, but they come to the surface in semi-satirical form in another trilogy of an invented world, Brian Aldiss's *Helliconia* novels. Helliconia is a planet populated by humans and various non- and quasi-human species. Close to civilized humans are the perpetually nomadic, pastoral but not-quite-human Madi. Somewhat further away are the monkey-like Nondads, and the even more animal-like 'Others' stand (or, rather, swing) at a more remote remove. Man, Madi, Nondads and Others make as unambiguous a man-from-the-apes scenario as any copywriter could wish for. The humans on Helliconia come to realize that they have inherited their world relatively recently from another species unrelated to humans, the fearsome Phagors, and begin to wonder about their own evolutionary origins. In *Helliconia Summer*, the second novel in the trilogy, scandal is caused when a human ruler takes a Madi princess for his queen, just as a scientist is trying to essay interbreeding experiments between Madi and Nondads, and Nondads and Others, to test his evolutionary theories.

[9] Dysfunctions which, incidentally, parallel the reactions of war veterans once they return home. One includes here that group of authors earlier mentioned which Shippey (in *J. R. R. Tolkien, Author of the Century*) has singled out as incapable of expressing the horror of their experiences in any mode other than the fantastic, as well as Frodo himself, who on returning to the Shire after his traumatic adventures, cannot observe his domestic scene in any way more involving than dispassionate, almost anthropological curiosity, and is therefore unable to settle down in Hobbit society again (*Rings* VI,9).

[10]*Ibid.* Again, in typical confusion, Tolkien continues to play with the idea that Orcs were corrupt Elves, and later, Men, perhaps with some Maiar mixed in to keep some kind of order amid this semi-human rabble. Still later ruminations suggest that Orcs were more in origins Men; later, Men could be reduced to Orc level and then mate with pre-existing Orcs. The 'last word' on Orcs was that they were derived from Men. However, this finality was imposed less by the author having settled the issue after

more than a half century of internal debate than by his death not long after these notes were composed.

Chapter 8: *The Last March of the Ents*

[1] See *HOME* VI for the earliest appearances of Ents in the context of the first drafts for what became *The Lord of the Rings*. Tolkien started 'A Long-Expected Party' sometime before December 19, 1937 (*Letters* 20), and by February 2, 1939 (*Letters* 35), had planned for the appearance of a giant, though his tree-ish connection was not fully realized. However, when we meet Treebeard himself as a character for the first time, rather than as a reference (a scrap of dialogue between Treebeard and Frodo on an abandoned letter dated June 27-29th, 1939) his arborescent nature is fully evident. A scrap of very late writing, 'Of the Ents and the Eagles,' is given and discussed in *HOME* XI.

[2] For a discussion on research into the consequences of inbreeding in cheetahs see R.M. May 'The cheetah controversy,' Nature 374, 309-310 (1995).

[3] Wiens, D. *et al.*, 'Developmental failure and loss of reproductive capacity in the rare palaeoendemic shrub,' *Dedeckera eurekensis, Nature* 338, 65-67, 1989.

[4] Tilman, D. *et al.*, Habitat destruction and the extinction debt, *Nature* 371, 65-66, 1994.

[5] See M. L. Smith, J.N. Bruhn and J.B. Anderson, 'The fungus *Armillaria bulbosa* is among the largest and oldest living organisms,' *Nature* 356, 428-431 (1992).

[6] See L. Simon and colleagues, 'Origin and diversification of endomycorrhizal fungi and coincidence with vascular land plants,' *Nature* 363, 67-69 (1993).

[7] See 'Net transfer of carbon between ectomycorrhizal tree species in the field' by Suzanne W. Simard and colleagues, *Nature* 388, 579-582 (1997).

Chapter 9: *O For the Wings of a Balrog*

[1] It is also reported that the Nazgûl-bird had a foul odor: in this context, it is interesting to report that the pterodactyls in Arthur Conan Doyle's *The Lost World* are notable for their unpleasant smell. Given Tolkien's

fondness for paleontology as well as fantasy fiction, it is a fair guess that he derived some of the Nazgûl-bird's characteristics from that source.

[2] See my book *In Search of Deep Time: Beyond the Fossil Record to a New History of Life* (Free Press, 1999) as well as the introduction to my book with artist Luis V. Rey, *A Field-Guide to Dinosaurs* (Barrons, 2003), for more detailed discussions of the process of paleontological myth-making.

[3] See Vizcaíno, S.F. and Fariña, R.A. 'On the flight capabilities and distribution of the giant Miocene bird *Argentavis magnificens* (Teratornithidae),' *Lethaia* 32, 271-278 (1999).

[4] See Bramwell, C.D. and Whitfield, G.R. 'Biomechanics of *Pteranodon,*' *Philosophical Transactions of the Royal Society of London* B 267, 503-581 (1974); and Brower, J.C. 'The aerodynamics of *Pteranodon* and *Nyctosaurus,* two large pterosaurs from the Upper Cretaceous of Kansas,' *Journal of vertebrate Paleontology* 3, 84-124 (1983).

Chapter 10: *Six Wheels on My Dragon*

[1] In his correspondence (*Letters* 19) Tolkien writes about having to lecture on dragons to the Natural History Museum in Oxford. Note that not all of Tolkien's dragons have wings. For most of the first age, the dragons were wingless. Glaurung, the dragon slain by Turin Turambar in *The Silmarillion,* had no wings and dragged himself around Beleriand on foot. The winged dragons made their first appearance in the legendarium with the the War of Wrath at the very end of the First Age, where the winged dragon Ancalagon is mentioned (*The Silmarillion,* chapter 24.)

[2] See 'Limbs generated at site of tail amputation in marbled balloon frog after vitamin A treatment' by P. Mohanty-Hejmadi, S.K. Dutta and P. Mahapatra, *Nature* 355, 352-353 (1992).

[3] See E.B. Lewis 'A gene complex controlling segmentation in *Drosophila,*' *Nature* 276, 565-570 (1978).

[4] See M.I. Coates and J.A. Clack, 'Polydactyly in the earliest known tetrapod limbs,' *Nature* 347, 66-69 (1990).

[5] See for example 'Tbx genes and limb identity in chick embryo development' by A. Isaac and colleagues (*Development* 125, 1867-1875, 1998) showing that the 'identity' of forelimbs and hindlimbs is controlled by particular genes in the Tbx gene family.

[6] Note that lampreys, although often called 'eels,' are incorrectly described

as such. True eels (*Anguilla*) are fish with jaws and paired fins.

[7]Biology is to be treasured for its exceptions, and there might just possibly be one naturally occurring six-legged vertebrate. The coelacanth (*Latimeria chalumnae*) is a peculiar fish distantly related to land vertebrates. Like land vertebrates, its paired pectoral (fore-) and pelvic (hind-) fins are supported by bones that are equivalent to our arms and legs. The resemblance was so close that the coelacanth was nicknamed 'Old Four-Legs.' The midline dorsal and tail fins, in contrast, are not supported by limb-like bones. However, the coelacanth has two other, median fins — the anal and second dorsal fins — that are supported by bones. Paleontologist Per Erik Ahlberg has suggested that these fins represent an instance of homeotic transformation that has become fixed in nature, in which two ordinary median fins have been replaced by leg-like fins ('Coelacanth fins and evolution,' *Nature* 358, 459, 1992.)

Chapter 11: *The Eyes of Legolas Greenleaf*

[1] See 'Chameleons use accommodation cues to judge distance' by L. Harkness, *Nature* 267:346 –349 (1977).

[2] See 'Scotopic color vision in nocturnal hawkmoths' by A. Kelber and colleagues, *Nature* 419, 922-925 (2002).

[3] See 'Richer color experience in observers with multiple photopigment opsin genes' by K.A. Jameson and colleagues, *Psychonomic Bulletin and Review* 8, 244-261 (2001).

[4] See 'Attraction of kestrels to vole scent marks visible in ultraviolet light' by Jussi Viitala and colleagues, *Nature* 373, 425-427 (1995).

Chapter 12: *Of Mithril*

[1] See 'A family of ductile intermetallic compounds' by K. Gschneidner, Jr. et al, *Nature Materials* 2, 587-590 (2003).

[2] See 'Possible high-Tc superconductivity in the Ba-La-Cu-O system ' by J. G. Bednorz and K. A. Muller, *Zeitschrift für Physik B: Condensed Matter* 64, 189-193 (1986).

[3] See 'Superconductivity above 130 K in the Hg-Ba-Ca-Cu-O system' by A. Schilling and colleagues, *Nature* 363, 56-58 (1993).

END-NOTES

Chapter 13: *The Laboratory of Fëanor*

[1] The greatness of Fëanor is mentioned briefly in *Rings* A, but his special nature is not there discussed. However, the ferocious spirit consumed his mother Miriel who, alone among the Elves, wished to die utterly, and the consequences of this are discussed in *HOME* X (see especially *Laws and Customs among the Eldar*).

[2] In an early draft of the relevant passage in *The Two Towers* (*HOME* VIII) Tolkien has the *palantír* shatter on impact, but he quickly changed his mind for obvious dramatic reasons. Another change of mind occurred in drafts for what became the *Rings* A, where it says that the *palantír* of Weathertop went down with King Arvedui in a ship wreck: earlier versions of the same passage say explicitly that the Stone was broken (*HOME* XII).

[3] For another example of the same phenomenon, see the episode in which Sam and Frodo attempt to cut their way through Shelob's webs (*Rings* IV,9): Frodo's High-Elven blade shears through cords that defeat Sam's lesser, Numenorean weapon.

[4] See for example 'Experimental quantum teleportation' by Dik Bouwmeester and colleagues, *Nature* 390, 575-579 (1997).

[5] See 'Formation of carbon nitride with sp^3-bonded carbon in CN/ZrN superlattice coatings' by M.L. Wu and colleagues, *Applied Physics Letters* 76, 2692-2694 (2000), and 'Materials science: Disciplines bound by pressure' by Paul F. McMillan, *Nature* 391, 539-540 (1998).

[6] Given that this book is about the science of Middle-earth, rather than the Blessed Realm, where Middle-earth rules need not apply.

Chapter 14: *The Gates of Minas Tirith*

[1] See Newton, S. *The Origins of Beowulf and the Pre-Viking Kingdom of East Anglia*, (Cambridge: D. S. Brewer, 1993).

[2] From *Beowulf in Hypertext*: http://www.humanities.mcmaster.ca/~beowulf/main.html.

[3] *Beowulf* lines 322-323, quoted in this context by S. Newton in *The Origins of Beowulf and the Pre-Viking Kingdom of East-Anglia* (Cambridge: D. S. Brewer, 1993).

[4] See K. Padian, 'A Daughter of the Soil: Themes of Deep Time and Evolution in Thomas Hardy's *Tess of the d'Urbevilles*,' *Thomas Hardy Journal* vol. 13, pp. 65-81, 1997. It is perhaps no coincidence that Padian is a paleontologist accustomed, unlike most literary critics, to thinking about intervals of time too great for regular mortals to comprehend.

[5] My daughter Rachel, then aged three, totally ignorant of Tolkien but maximally receptive to images of fairy-tale princesses, saw the merest glimpse of Liv Tyler as Arwen in a trailer to Peter Jackson's *The Two Towers*, and immediately exclaimed that she looked just like Snow White — an observation which, given Ms. Tyler's raven hair and pale complexion, seems absolutely accurate. It struck me that this resemblance makes perfect sense according to the thesis that the characters of conventional fairy-tale had their origin in the real, heroic figures of a lost past that Tolkien was trying to reconstruct in fiction. That is, the tale of Snow White and the Seven Dwarves could be seen as the ultimate bowdlerization of *The Lord of the Rings* in the same way that *Hey Diddle Diddle* is the last vestige of Frodo's poem about the Man in the Moon. Aragorn becomes the nameless Prince Charming; the evil Queen stands for Sauron, or even Galadriel (in which case the whole business of the effects of the poisoned apple — perhaps an allegory of The One Ring — becomes even more fascinating) and the supporting cast of dwarves and assorted woodland creatures stands for everyone else.

[6] The word 'Gnomes' for Noldor has entirely disappeared from *The Lord of the Rings* and *The Silmarillion* but it is ubiquitous in earlier versions of the mythology, as can be seen (for example) in *HOME* I. Tolkien himself writes on confusion attendant on the word *gnome* in *Letters* 239.

[7] See 'Hominid revelations from Chad' by Bernard Wood, *Nature* 418, 133-135, 2004.

[8] See 'Latest *Homo erectus* of Java: potential contemporaneity with *Homo sapiens* in Southeast Asia' by C.C. Swisher III and colleagues, *Science* 274, 1870-1874 (1996).

Chapter 15: *The Lives of the Elves*

[1] Théoden is born in 2948 of the Third Age and dies in battle on the Pelennor Fields on March 15, 3019, aged 71. Aragorn is born on March 1, 2931 and dies aged 210 in the year 1541 of the Shire, equivalent to 3141 of

the Third Age. Interestingly, the steward Denethor II was born in 2930, making him Aragorn's senior by just one year (*Rings* B 'The Tale of Years').

[2] See 'The cultural wealth of nations' by Mark Pagel and Ruth Mace (*Nature* 428, 275-278, 2004) and references therein.

[3] The classic paper 'Mitochondrial DNA and human evolution' by Rebecca L. Cann, Mark Stoneking, and Allan C. Wilson (*Nature* 325, 31-36, 1987) introduced the concept of a single, recent African ancestry of modern humans in almost biblical terms.

[4] See 'Growth processes in teeth distinguish modern humans from *Homo erectus* and earlier hominins' by C. Dean and colleagues, *Nature* 414, 628-631 (2001).

[5] Melov, S. et al., Extension of life-span with superoxide dismutase/catalase mimetics, *Science* 289, 1567-1569 (2000).

[6] Migliaccio, E. et al. The p66[shc] adaptor protein controls oxidative stress response and life span in mammals, *Nature* 402, 309-313 (1999).

[7] A curious exception is Círdan the Shipwright, who is described as old and grey (*Rings* VI,9). I cannot account for this in any way apart from the symbolic, in that Círdan fulfils the role of Charon, the boatman of the Styx — and with Charon in mind, Círdan looks rather perky by comparison.

[8] Arwen was born in 241 of the Third Age (*Rings*, B). The *Annals of Aman* (in *HOME* X) places Galadriel's birth in the Valian Year 1362, each of which is approximately equal to ten years of the Sun. The first Sun rose in Valian Year 1500, or 15,000-13,620 = 1,380 solar years later. The First Age lasted another 600 years, and with the combined total of the Second and Third Ages of 6,460 years, Galadriel when we see her is 8,440 years old.

Chapter 16: *Giant Spiders and 'Mammoth' Oliphaunts*

[1] For an accessible introduction to the latest controversies on the relationships between size, shape, and metabolism, see John Whitfield's article 'All Creatures Great and Small,' *Nature* 413, 324-344 (2001).

[2] See 'Mammoths in miniature' by A.M. Lister (*Nature* 362, 288-289, 1993).

[3] See M. Fortelius and J. Kappelman, *Zoological Journal of the Linnean Society* 108, 85-101 (1993).

[4]See for example Robert Bakker's essay 'Return of the Dancing Dinosaurs,' p. 38-69 in vol, 1 of *Dinosaurs Past and Present* (ed. S.J. Czerkas and E.C. Olson, Los Angeles: Natural History Museum of Los Angeles County in association with the University of Washington Press, 2 vols., 1987)

Chapter 17: *Indistingushable from Magic*

[1] Clarke, A.C., 'Technology and the Future,' in *Report on Planet Three and Other Speculations* (London: HarperCollins, 1972).

[2] The transformation of technology into myth and magic is seen very well in that sub-genre of science fiction known as 'post-apocalyptic' or 'post-holocaust' that flourished during the Cold War in the decades immediately after World War II. One thinks of Walter Miller's *A Canticle for Leibowitz* (1960), in which a monk in post-nuclear Texas uses electronic circuit-board diagrams as the basis for an illuminated manuscript, his lack of comprehension similar, one might imagine, to those monks who copied Aristotle or *Beowulf*. Another example is found in George R. Stewart's *Earth Abides* (1949), in which the children of a commune in a disease-depopulated America turn away from the adults' attempts to instill an American education, and retreat into a self-created mythology. It might be too much of a stretch to place *The Lord of the Rings* alongside such works, but they emanate from the same period, and are imbued with the same sense of pessimism, regret, and longing for vanished, better days now long past.

[3] In his book *The Hacker Crackdown: Law and Disorder on the Electronic Frontier* (Penguin, 1994) a history of telephony, Bruce Sterling tells what happened when a telephone came to the telegraph office of a small Midwestern town. The whole town turned out to admire this device, and the mayor, in a fit of prolepsis, declared that, someday, *every* town in America would have one.

[4] For a good overview of this field, see the article 'Actions from Thoughts' by Miguel A. L. Nicolelis, *Nature* 409, 403-407 (2001).

[5] This kind of technological utopia forms the basis of Iain M. Banks's brilliantly realized galactic civilization known simply as 'The Culture,' which forms the backdrop to several novels, notably *Consider Phlebas* and *The Player of Games*. In these novels, Banks considers how a utopian society might respond to threats imposed by more primitive societies

that are driven by the kinds of passionate, religious urges that the Culture has long abandoned. The answer is simple: after a certain amount of soul-searching, the Culture either absorbs the enemy or crushes it mercilessly.

[6] *Rings* VI,2; Tolkien notes that Frodo and Sam, in Mordor, know nothing of the slave-worked fields around the Sea of Nurnen, the 'breadbasket' of Mordor of which the Plateau of Gorgoroth was the industrial heartland.

[7] In *Rings* IV,1, Sam is startled that a well-knotted *hithlain* rope came loose at a single tug, after it had just borne the weight of himself and Frodo. In this way, the weave of *hithlain* has much in common with Frodo's mail shirt — that it responds smoothly to continuous pressure, but stiffens up when required to resist sudden shocks. For more on this, see chapter 12, 'Of *Mithril*.'

[8] See 'An artificial landscape-scale fishery in the Bolivian Amazon' by Clark Erickson, *Nature* 408, 190-193 (2000).

[9] For an overview of GFP, see 'The Green Fluorescent Protein' by Roger Y. Tsien, *Annual Reviews of Biochemistry* 67, 509-544 (1998).

[10] In Lothlórien (*Rings* II,6), Frodo has the strange sense that he is looking on a vanished world from a high window. This, for literary effect, reminds me very strongly of Keats' *Ode to a Nightingale*, in which the poet reports the effect of the nightingale's song transporting him to

> Charm'd magic casements, opening on the foam
> Of perilous seas, in faery lands forlorn.

This poem resonates with much of Tolkien's earlier poetry, a topic which I do not have space to discuss here: however, like much of the material on the Elves, it deals with the pains of mortality, contrasting it with immortality. The same passage (*Rings* II,6) deals playfully with two other scenarios of temporal distortion or bifurcation: Frodo is reported as staying in Lothlórien even after he had, in fact, left, and he is separately described as having the impression of being physically in a remote period of time long past, when the seabirds he can hear have since become extinct.

[11] *Rings* II,1. Bilbo also notes that in Rivendell there is evidence for distortions of space as well as time. Although Rivendell is a big house, he says, there is always more of it to discover. This suggests that there is something more than three-dimensional to the Last Homely House, a

common theme in science fiction (see, for example, Robert Heinlein's story *And He Built a Crooked House*), and an all-pervading one in the engravings of M.C. Escher, whose interiors always remind me of Rivendell.
[12] See 'Light speed reduction to 17 metres per second in an ultracold atomic gas' by L. V. Hau and colleagues, *Nature* 397, 594-598 (1999), and 'Observation of coherent optical information storage in an atomic medium using halted light pulses' by C. Liu and colleagues, *Nature* 409, 490-493 (2001).
[13] *Rings* VI, 3: actually, not quite: the light from the star-glass appears once more, only to be extinguished, as Frodo goes into the West at the very end of the book. This would, of course, make the speed of light in the glass even slower.
[14] A similar case of water associated with time-shifting is that of the faces of warriors killed in battles long ago, glimpsed in the pools of the Dead Marshes (*Rings* IV,2), although in that instance the cause presumably owes less to Silverlode than sorcery.

Chapter 18: *In the Matter of Roots*

[1] *The New Concise British Flora* by W. Keble Martin (London: Mermaid, Michael Joseph, 1986) has been invaluable in the research for this chapter.
[2] The fir is mentioned more often than any other tree in *The Lord of the Rings*. I have found six references (*Rings* I,3; I,12; II,6; II,9; III,4; IV,1): importantly in this context, neither the Douglas fir *Pseudotsuga* nor any other species of fir (such as the Silver fir, *Abies*) are native to the modern British flora. However, their fossil pollen has been recovered from various deposits from the past two million years or so, so they might be regarded as British in a very broad sense.
[3] The big, white water-lily familiar from ponds and Monet paintings is *Nymphaea alba*, and this is the species gathered by Tom Bombadil for Goldberry (*Rings* I,6).
[4] 'Cornel' is an archaic word and could refer either to the 'cornelian cherry,' a fruiting tree *Cornus mascula*, native to southern Europe; or to dogwood (*C. sanguinea*), a shrub that grows in southern England.
[5] See *Letters* 312 for speculations about the kinship of some of Tolkien's imagined flora. In *HOME* VII, *niphredil* and *simbelmÿne* are both compared with the snowdrop (*Galanthus nivalis*).

[6] See 'Evolution, consequences and future of plant and animal domestication' by Jared Diamond, *Nature* 418, 700-707 (2002).

[7] The incongruity of potatoes is matched by that of the rabbits that Samwise is cooking, as noted by Shippey in *The Road to Middle-earth*: rabbits, like potatoes, may seem so much a part of the English scene that you'd be forgiven for thinking that they had always been there — when they are, in fact, an introduced species.

[8] See 'Tobacco to tomatoes: a phylogenetic perspective on fruit diversity in the Solanaceae' by Sandra Knapp, *Journal of Experimental Botany* 53, 2001-2022, (2002).

[9] See A. J. Stuart, 'Pleistocene occurrences of the European pond tortoise (*Emys orbicularis* L.) in Britain,' *Boreas* 8, 359-371 (1979).

[10] See *Frozen Fauna of the Mammoth Steppe: The Story of Blue Babe* by R. Dale Guthrie (Chicago: University of Chicago Press, 1990) for a lively account of the debate.

Chapter 19: *The One Ring*

[1] See *Figments of Reality: The Evolution of the Curious Mind* by Jack Cohen and Ian Stewart (Cambridge; Cambridge University Press, 1997) for an excellent and provocative look at how the mind works.

[2] See *Laws and Customs among the Eldar* in *HOME* X. Turning for a moment from theoretical existentialism to experimental philology, I have wondered whether the word *hröa* is in any way related to the word *hnau*, used by C. S. Lewis in his *Out of the Silent Planet* sequence as word to denote sentience (to which Tolkien referred directly in *The Notion Club Papers* in *HOME* IX), or to the word *hrön* invented by Jorge Luis Borges in his story *Tlön, Uqbar, Orbis Tertius*, meaning an object from an imaginary world that appears in everyday reality.

[3] See 'The Search for Extra Dimensions' by Steven Abel and John March-Russell in the November 2000 issue of *Physics World*. This gives an accessible account of the very latest in multidimensional physics without which I could not possibly have approached the subject.

[4] The effects on immortals are, presumably, different, given that Elves are said to be able to experience the real and spirit-worlds simultaneously (*Rings* II,1).

[5] Susumu Tachi at the University of Tokyo has recently demonstrated an invisibility cloak, in which a fabric consisting of 'retro-reflective spheres' broadcasts an image of a scene the other side of the wearer, making the wearer and indeed anything the wearer might be hiding appear transparent (see Tachi, S., 'Telexistence and retro-reflective projection technology (RPT),' Proceedings of the 5th Virtual Reality International Conference (VRIC2003), pp.69/1-69/9, Laval Virtual, France, May 13-18, 2003.) This ingenious item of 'smart' clothing does not solve the problem of the Ring, however: even were the Ring able to exert its influence over clothing, the clothing itself would need to be 'primed' in some way. As far as we know, the Ring can make anything invisible, so the problem of invisibility remains.

[6] See 'Yttrium and lanthanum hydride films with switchable optical properties' by J.N. Huiberts and colleagues, Nature 380, 231-234 (1996).

[7] With the possible exception of the sword Sting, whose blue glow in the presence of Orcs Samwise believes can be seen, even when the bearer is wearing the Ring (Rings IV,10).

Chapter 20: *Science and Fantasy*

[1] There is a philological connection between science and fantasy that stretches back beyond recorded history. In *Splintered Light*, Verlyn Flieger compares the words 'phenomenon' and 'fantasy,' derived respectively from the Greek verbs *phainesthai* ('to appear') and *phantazein* ('to make visible'). These are, in turn, intransitive and transitive elaborations on an earlier verb *phainein* ('to show'). All three forms can be traced to a yet earlier Indo-European root form *bhā-* ('to shine'). A distinct root — which, however, has the same sound — means 'to speak,' and is the source of Greek words such as *phōnē* (a sound) and *phonēin* ('to speak'). Long ago, says Flieger, light, language, knowledge, and revelation expressed much the same things — science and fantasy would have been equivalent concepts.

[2] Richard Dawkins, 'Universal biology' (book review of *Artificial Life: The Quest for a New Creation* by Steven Levy), Nature 360, 25-26 (1992).

[3] In *Splintered Light*, Verlyn Flieger suggests that Tolkien's frequent use of towers from which one might glimpse far countries or the sea was an

expression of an attribute of the human psyche, that is, the desire to seek something without knowing what it is.

[4] Durant, J. 'Pseudo-science, complete fiction,' The *Independent*, August 21, 1998 (p. 13); see the response in *Nature*, August 28, 1998.

[5] The lengths to which the production team went to ensure that the science-lab backgrounds of *The X-Files* looked real are described by Anne Simon in her book *The Real Science Behind the X-Files: Microbes, Meteorites and Mutants* (Simon and Schuster, 1999). See my review in *Nature* 403, 135-136 (2000). In *Splintered Light*, Verlyn Flieger, commenting on Tolkien's essay *On Fairy Stories*, suggests that one purpose of the fantastic is to allow us to look at familiar things in a new light — a shift of perspective that is essential to scientific thought.

INDEX

INDEX

INDEX

ABOUT THE AUTHOR

Henry Gee was born in 1962. He received his bachelor of science degree in zoology and genetics at the University of Leeds in 1984, and his doctorate in zoology at the University of Cambridge in 1991. In 1987 he joined the staff of *Nature*, the leading international journal of science, where he is now Senior Editor, Biological Sciences. In 1996 he was awarded a Regents' Professorship of the University of California. He is the author of several books including *Jacob's Ladder* (Norton, 2004), *In Search of Deep Time* (Free Press, 1999) and (with Luis V. Rey) *A Field Guide to Dinosaurs* (Barrons, 2003). As *Olog-Hai* he is the occasional science correspondent for TheOneRing.net, the leading Tolkien fan website. He lives in Ilford, Essex, England. Learn more about Henry and his current projects at his website, www.chiswick.demon.co.uk.

Cold Spring Press

An Imprint of Open Road Publishing
P.O. Box 284
Cold Spring Harbor, NY 11724
Jopenroad@aol.com

Nonfiction:

The Science of Middle-earth, by Henry Gee, $14.00

More People's Guide to J.R.R. Tolkien, by TheOneRing.net, $14.00

The People's Guide to J.R.R. Tolkien, by TheOneRing.net, $16.95

The Tolkien Fan's Medieval Reader, by Turgon, $14.95

Tolkien in the Land of Heroes, by Anne C. Petty, $16.95

Dragons of Fantasy, by Anne C. Petty, $14.95

Myth & Middle-earth, by Leslie Ellen Jones, $14.95

The 100 Best Writers of Fantasy and Horror, by Douglas A. Anderson, $16.95 (Spring 2005)

Fiction:

The Sillymarillion, by D.R. Lloyd, $11.00

Book of the Three Dragons, by Kenneth Morris, $11.95

Lud-in-the-Mist, by Hope Mirrlees, $11.00 (Spring 2005)

Thin Line Between: Book One of the Wandjina Quartet, by M.A.C. Petty, $11.00 (Spring 2005)

For US orders, include $5.00 for postage and handling for the first book ordered; for each additional book, add $1.00. Orders outside US, inquire first about shipping charges (money order payable in US dollars on US banks only for overseas shipments). We also offer bulk discounts. *Note: Checks or money orders must be made out to* **Open Road Publishing**.